Aesthetics of Early Sound Film

Aesthetics of Early Sound Film

Media Change around 1930

Edited by
Daniel Wiegand

Routledge
Taylor & Francis Group
LONDON AND NEW YORK

Cinema and Technology is published in association with the Laboratoire CinéMédias at Université de Montréal and the TECHNÈS International Research Partnership on Cinema Technology, supported by The Social Sciences and Humanities Research Council of Canada Partnership Grant, as well as The Canada Research Chair in Cinema and Media Studies.

The publication of this book is made possible by a starting grant from University of Zurich.

First published in 2023 by Amsterdam University Press Ltd.

Published 2025 by Routledge
4 Park Square, Milton Park, Abingdon, Oxon OX14 4RN
605 Third Avenue, New York, NY 10158

Routledge is an imprint of the Taylor & Francis Group, an informa business

ISBN: 9789463727372 (hbk)
ISBN: 9781041175292 (pbk)
ISBN: 9781003690481 (ebk)
NUR 670

Cover illustration: Greta Garbo and Clarence Brown in a set photograph for *Anna Christie* (1930) (courtesy of Wisconsin Center for Film and Theater Research)

Cover design: Coördesign, Leiden

DOI: 10.5117/9789463727372

Table of Contents

IV Speech and Language

V Music and Noise

1. The Aesthetics of Early Sound Film: An Introduction

Daniel Wiegand

Abstract: This introduction outlines the emerging field of early sound film studies, arguing that the transitional era around 1930 should be conceived in terms of a specific aesthetics characterized by gradual processes of renegotiation and reorientation. Rather than a period of aesthetic restrictions, it is one of aesthetic options and experimentation, less a unidirectional break than a protean and polymorphous period, which is embedded in film history in complex ways. In addition, more global and transnational perspectives on the media change are needed, along with increased visibility of early sound films, including those often marginalized in scholarship.

Keywords: media change, film history, early sound film, aesthetics

"What is this and what does it mean for us?" These might be two of the questions running through Greta Garbo's and Clarence Brown's heads as they suspiciously, but somewhat benevolently, look up at the microphone suspended above them in a 1930 set photograph for *Anna Christie*, Garbo's first talkie, which was directed by Brown (see cover image). Looking at the picture today, we might be reminded of our own encounters with new technologies and apparatuses, some of them quite recent, others already a little older: holding a smart phone in our hands, scrolling down a website for the first time; or looking at a video conference 'set-up' on our computer screen, with a small mirror image of ourselves next to the other participants.

New technologies are usually greeted with a mixture of scepticism, hopeful expectation, dreams and musings about their possible futures, and even more importantly, about *our* futures *with* them. When images and sounds

Wiegand, D. (ed.), *Aesthetics of Early Sound Film: Media Change around 1930*.
Taylor & Francis Group, 2023
DOI 10.5117/9789463727372_CH01

are involved, especially in the realm of art and entertainment, aesthetic issues necessarily play a part in these transformations and projections into the future. Thus, the large-scale introduction of sound technologies into the film industry around 1930 resulted in major changes for film aesthetics, when the look and sound of films and the ways they were perceived by spectators were undergoing profound shifts. Take the beginning of *Die Nacht gehört uns* (*The Night Belongs to Us*) (1929, dir. Carl Froelich), one of the very first 'all-talking pictures' produced in Germany and, presumably, for many spectators in Europe the first feature film with sound they ever saw. After the music for the opening credits has faded away, there is a brief close-up of one of the characters, company boss Marten, played by Walter Janssen, who puts a cigar into his mouth and inhales (Fig. 1.1); next is a long shot, in which Marten stands in his office, blows out the smoke (Fig. 1.2), then abruptly turns around and walks away from the camera. During the first shot and the beginning of the second, no recorded sound can be heard, just the obtrusive ground noise *resulting* from the recording.[1] Then, a faint sound is audible, which could be the actor's exhaling breath, followed by sounds that are more clearly identifiable as footsteps, with their reverberation indicating that they were most likely recorded on the set.

Clearly, the addition of synchronized sound alters this sequence of shots and its potential perceptions in crucial ways. Imagine sitting in a cinema in 1929, full of anticipation of watching a film with sound, a *talkie*. During the credits, you listen to the recorded score coming from the loudspeakers recently installed in the auditorium – interesting, perhaps, but a far cry from the voluminous sound of the live orchestra you are used to. Then the music fades away and you hear ... not talk but *silence*; you look at the man's close-up and listen attentively; when he inhales inaudibly, you might hear your own breath of expectation, or your attention might be drawn to the ground noise coming from the speakers, which obtrusively foregrounds the new sound technology even at the moment of silence.[2] Finally, you hear the character's breath and his footsteps, in perfect sync with his movements. The effect of lifelikeness is astonishing, and you may feel even more immersed into this now sonorous world on-screen, the sounds filling the auditorium and mixing with your own sounds and silence.

1 On the problem of ground noise and the development of the first noise reduction systems in Germany, see Müller, *Vom Stummfilm zum Tonfilm*, 208–12. For the USA, see Jacobs, "The Innovation of Re-Recording."

2 On the relation of early sound films and acoustic silence, see O'Rawe, "The Great Secret: Silence, Cinema and Modernism," Moure: "Du silence au cinema," and my own Wiegand, "The Delightful Paradox" (forthcoming).

Figure 1.1: Walter Janssen as company boss Marten draws on his cigar.

Figure 1.2: Marten blows out the smoke.

The emergence of synchronized sound in global cinema around 1930 and throughout the 1930s introduced new forms of audience address as well as new potentials for film style, and, more specifically, for the juxtaposition of images and sounds – be it dialogue, music, background noises, or silence. As Béla Balázs optimistically remarked in the midst of the period: "Technical innovation is the most effective inspiration. It is the muse itself."[3] However, the specific ways in which these new potentials should and would unfold were by no means obvious. Rather, the future of sound film aesthetics was still undefined and subject to constant renegotiation and redefinition in written discourse as well as in practice. What precisely *was* sound film, or rather, what *could* it be? Filmmakers, audiences, exhibitors, technicians, critics, and theorists worldwide were faced with questions such as: how much dialogue should be in sound films? Should dialogue and music be heard at the same time? How does sound alter the 'nature' of film and its status as an art form? Which formal features of silent film should and could be continued? In all the major film-producing countries we find a wealth of writings that reflect this "crisis of transition"[4] and document the manifold debates about the future of sound film and its aesthetics, and research has only partially reappraised and explored this rich body of work.[5] Thus, even though the conversion happened significantly faster than other media changes in film history (at least in some countries), these debates show that the coming of sound was neither a sudden rupture nor a pre-planned and linear transition but rather a gradual process of renegotiation and reorientation.

This volume wishes to draw attention to the various aesthetics emerging internationally in this protean and polymorphous period that we often refer to as 'early sound film.'[6] While the era has often been reduced to the status of a mere pathway into 'classical cinema,' no more than a brief interruption between two more consolidated phases – the 1920s and the 1930s – one aim of the volume is to look at early sound film as a distinct phase

3 Balázs, *Early Film Theory*, 184.

4 Balázs uses this term as early as 1930 (ibid., 207). For a use of the term "crisis historiography," also in relation to early sound film, see Altman, *Silent Film Sound*, 15–23; and Wedel, *Pictorial Affects, Senses of Rupture*.

5 English translations and discussions of source texts from the sound film debates can be found in Kaes, Baer, and Cowan, *The Promise of Cinema*, 549–755 (Germany); Abel, *French Film Theory and Criticism* (France); Weis and Belton, *Film Sound: Theory in Practice*. Additional studies dealing with the aesthetic reception of early sound film include Crafton, *The Talkies* (USA); Szczepanik, "Sonic Imagination" (Czechia); Porter, "Okay for Sound?" (UK); Mühl-Benninghaus, *Das Ringen um den Tonfilm* (Germany); Wiegand, "Islands of Sound in the Silent Flow of Film" (Germany).

6 Of course, sound films were produced even earlier than that, e.g., Oskar Messter's Tonbilder. The term here refers to the media shift around 1930 that affected the film industry at large.

in film history, which merits scholarly investigation in its own right. The individual chapters in this book take a closer look at films from the period to find that aesthetic practices were more heterogeneous than has often been assumed. Such an endeavour, I feel, is still needed. Some preconceived assumptions about early sound film aesthetics – such as the predominance of static cameras and stilted acting – are quite persistent, and while these are certainly true to some extent, they do not tell the whole story. *The Night Belongs to Us* is a case in point: the opening sequence is a bravura piece of fast crosscutting that uses the various voices and sound devices present in the scene (e.g., a telephone, loudspeakers, a switch board, headphones) to connect the different shots and localities, thereby self-consciously displaying its fascination with modern technology and speed (the scene shows a car race). The entire film reveals a strong interest in the use of sound effects, linguistic diversity, and partly unintelligible dialogue, which – rather than being 'stilted' and 'theatrical' – betrays a shift towards everyday realism.[7] The film thus reveals that, rather than being merely inhibited by technology, filmmakers in the early sound film period explored the new resources offered by synchronized sound from the outset.

As this volume seeks to flesh out, then, through an array of in-depth case studies, the early sound film era should be conceived less as a period of aesthetic *restrictions* than one of aesthetic *options* and *experimentation* – a richness that is most apparent at the intersection of historical research and formal analysis. Understood as a period of 'fruitful uncertainty,'[8] the years of the conversion to sound are, to some extent, comparable to the years of early cinema around 1900, another period of experimentation and fundamental change. Thus, when several commentators during the early sound film period feared that synchronized sound would lead the cinema back to its 'primitive' origins, this was perhaps somewhat true; in a sense, early sound film *was* a return, only not to some kind of supposed 'primitivity' but to the openness and plurality that we can still sense in many films from the time before cinema became institutionalized around 1910.[9] That cinema was also ripe

7 For an analysis of the opening sequence, see Wiegand, "Listening to Faint Sounds and Silence." Jessica Berry discusses this film's reception in Switzerland; see her article in this volume. For its reception in Germany, see Wiegand, "Entdeckungsfahrt in die Welt der Geräusche."

8 I am taking my cue here from Donald Crafton's notion of the "uncertainty of sound" (*The Talkies*, 1–18).

9 From the multitude of texts dedicated to early cinema studies, let me just mention Lewinsky, "The Best Years of Film History" and Gunning, "From the Bottom of the Sea" from the same volume, both of which stress the "aesthetic and narrative possibilities" and the "'anything can happen' aspect of early cinema" in relation to film programming (Gunning, 39; Lewinsky, 25).

with the infamous "dead ends"[10] that media archaeology has so persistently striven to unearth, and which we find abundantly in the early sound film period, too. Some of these 'unpursued paths' have only received scholarly attention recently, for instance, the various hybrid formats between silence and sound, among them part-talkies and so-called "sound versions," which were a widespread phenomenon in 1930s Japanese cinema.[11]

Another parallel that we might draw between early cinema and early sound film is the abundance of carryovers from earlier media practices and intermedial exchange in general. As David Thorburn and Henry Jenkins have suggested, phases of media change often develop an "aesthetics of transition,"[12] which is usually characterized by "impulse[s] of continuity" and "holdovers of old practices and assumptions" but also by heightened forms of "self-reflexivity and imitation" of other media.[13] As some of the chapters in this volume demonstrate, these overlaps can be observed in the early sound film period. Many conversion-era films were 'infiltrated' by expressive devices usually associated with silent film, while others emphatically displayed their connections to other sound media such as radio or the emerging record industry.[14]

If I agree, then, with Martin Barnier that the early sound film period as a whole constitutes a heterogeneous field of aesthetic experimentation that cannot be fully integrated into a seamless evolution towards "classical cinema,"[15] we nevertheless find in many early sound films the beginnings of an aesthetics that would emerge more fully in the consolidated era of the late 1930s. In fact, several chapters in this volume point to such 'germs.' In sum, however, rather than reflecting a unidirectional 'change' or 'break,' the corpus of early sound films is embedded in film history in more complex ways, with films often consisting of several layers and varying speeds and directions, betraying elements of both continuity and disruption, simultaneously pointing backwards and forwards, sometimes into several periods at once. For instance, part of the perceived newness of *The Night Belongs to Us* lay in its radical renunciation of the type of musical accompaniment that had characterized 1920s silent film screenings (and even some early

10 Huhtamo and Parikka, *Media Archaeology*, 3.

11 Nordström, "Between Silence and Sound"; Wiegand, "Islands of Sound." See also Johan Nordström's chapter in this volume.

12 Thorburn and Jenkins, *Rethinking Media Change: The Aesthetics of Transition.*

13 Ibid., 7, 10.

14 For early sound film's intermedial connections, see e.g., Crafton, *The Talkies*; Wurtzler, *Electric Sounds.*

15 Barnier, *En route vers le parlant*, 215.

sound film efforts) in Germany. This 'disruption' led the way for German talkies for the next few years and came to be associated with modernity and realism. However, nondiegetic music and dialogue underscoring increased steadily over the next few years, so that from a later perspective, many early talkies seem unusually quiet and even 'cold.' But if relinquishing the musical score seems like a 'dead end' with regard to 'classical' cinema, the trend reappears at later historical moments, for instance, in some of the emerging new wave cinemas. Thus, the sparse musical style of early sound films should be seen neither as *the* future of sound film (as often argued in contemporary discourse) nor as a dead end but more like an access point connecting to several other moments in film history, be it by similarity, congruence, or sharp contrast.

While this book attempts to take a fresh look at the aesthetics of early sound film, all the chapters necessarily draw on substantial research from the past decades, even though 'early sound film studies' as a clearly demarcated, international research field does not seem to exist yet. If studies of the transition from silence to sound have often focused on technological and economic developments,[16] aesthetic implications of the media change were also studied early on – in English, most notably in the works of Kristin Thompson, David Bordwell, and Barry Salt, often within the framework of a historiography of film style.[17] Recent years have seen a notable increase in book-length studies in English, dedicated, at least in part, to the aesthetics of early sound film, focusing on a range of specific aspects, such as music and songs,[18] the voice,[19] rhythm,[20] reception aesthetics,[21] sound and colour,[22]

16 Standard works in the field include Crafton, *The Talkies*; Gomery, *The Coming of Sound* (USA); Müller: *Vom Stummfilm zum Tonfilm* (Germany); Barnier, *En route vers le parlant* (France); Christie, "Making Sense of Early Soviet Sound"; Iwamoto, "Sound in the Early Japanese Talkies." More recently, an edited volume has brought increased attention to the transition in Japan for an English-language readership: Raine and Nordström, *The Culture of the Sound Image in Prewar Japan*.

17 Among the pioneering texts are Wood, "Towards a Semiotics of the Transition to Sound"; Thompson, "Early Sound Counterpoint"; many of the articles in Weis and Belton, *Film Sound: Theory in Practice*; as well as sections in Salt, *Film Style & Technology*; Bordwell, Thompson and Staiger, *The Classical Hollywood Cinema*; and Weis, *The Silent Scream*.

18 Spring, *Saying It with Songs*; Fleeger, *Sounding American*; Slowik, *After the Silents*; Lewis, *French Musical Culture and the Coming of Sound Cinema*; Wedel, *Pictorial Affects, Senses of Rupture*; O'Brien, *Movies, Songs, and Electric Sound*.

19 Kaganovsky, *The Voice of Technology*.

20 Jacobs, *Film Rhythm After Sound*.

21 Spadoni, *Uncanny Bodies*.

22 Street and Yumibe, *Chromatic Modernity*.

and sound engineering.[23] Some recent edited volumes and special issues of journals as well as individual articles have examined specific national contexts.[24] Additionally, the growing number of books on the cinema's aural dimension in general has also contributed to our understanding of early sound film.[25] The idea for the present volume grew precisely out of the wish to bring together some of this recent scholarship and to thereby increase awareness of early sound film studies as a growing international research field.

Not least for practical reasons, early sound film has often been studied with regard to specific national contexts, while comparative, global, and transnational perspectives are rare.[26] It is one of the premises of this volume, however, that such perspectives are necessary if we want to achieve a broader understanding of the global cinema's shift to sound. *Aesthetics of Early Sound Film* is therefore not limited to one national context but assembles studies on a range of different national cinemas, including the Soviet Union, Japan, the USA, Germany, France, Italy, the UK, and Switzerland. It goes without saying that this scope remains limited and needs to be broadened further in the future. Moreover, taking this perspective bears risks of its own. For instance, the period that one could define as 'early sound film' differs from country to country, or region to region, in some cases even considerably.[27] Therefore, looking at the transition from silence to sound globally should not be seen as an act of homogenization but, on the contrary, of sharpening awareness for differences, similarities, and specificities.

23 Hanson, *Hollywood Soundscapes*.

24 Davidson and Rippey, *Early Sound Cinema in the Late Weimar Republic*; Helmers, "The Transition from Silent into the Sound Era."

25 Some examples are Cooke and Ford, *The Cambridge Companion to Film Music*; Buhler, Neumeyer and Deemer, *Hearing the Movies*; Nasta and Huvelle, *New Perspectives in Sound Studies*; Beck and Grajeda, *Lowering the Boom*.

26 Notable exceptions are O'Brien, *Cinema's Conversion to Sound* and *Movies, Songs, and Electric Sound*.

27 Even for each national context, it is quite difficult to define when the early sound film period begins and when it ends. Nonetheless, many scholars have felt an urgency to do so and have proposed specific dates, such as 1926–1931 for the USA (Crafton, *The Talkies*) or 1926–1934 for France (Barnier, *En route vers le parlant*). Most often, such periodization is based on criteria relating to film production (e.g., the numbers of silent films still produced), specific technologies (e.g., widespread use of rerecording or noise reduction systems), and exhibition practices (e.g., numbers of theatres wired for sound). However, the proposed periodization is often based on fiction feature films, whereas the inclusion of other genres and formats such as documentaries or short films complicates matters further. Finally, one could ask if the period should perhaps be divided into even shorter phases. In Hollywood, for instance, 1926–1928 was arguably a completely different phase than 1929–1930.

Regardless of national context, films from the tranitional period are often marginalized, in research as well as in public discourse. Except for a few 'classics' that have made the leap into the canon of film history, relatively few films from the transitional period are shown in cinemas, re-released, or discussed in scholarship. One reason for this is their lack of availability and often poor state of preservation. As with many silent films, some of the earliest and historically most important sound films have to be considered lost, including many part-talkies. Seen as mere 'test runs' from the outset, many of these transitional films were never properly archived, yet these are precisely the works that could help shed light on the media change today.[28] The neglect of early sound films continues to this day in the general reluctance to properly restore and make accessible existing prints of films that are often deemed uninteresting or tedious for a modern audience – even one that is cinephile and historically informed. International collectors' circles, now increasingly active on the internet, sometimes offer access to films, albeit often at low quality or in otherwise problematic versions (for instance, when their provenance is unclear). That said, the situation *is* improving, with more films from the early sound film period becoming officially available[29] and being screened at international festivals.[30] It is one aim of this volume to contribute to this growing visibility of early sound films, by drawing attention to their aesthetic specificities and their complex imbrications in one of the most profound transformations in the history of the cinema.

The fifteen chapters in this volume are grouped together in five thematic sections. Section I, "From Silence to Sound," looks at transitions from silent to sound film and the aesthetic challenges involved by focusing on three distinct phenomena: the staging of dialogue scenes, the deployment of expressive visual effects, and animation's shift from the use of icons and picture-words to that of synchronized music and sound effects.

Through a close reading of several American transitional films, in particular John Ford's *Arrowsmith* (1931), Lea Jacobs traces the filming of dialogue scenes from the late silent into the early sound era, when multiple-camera

28 An example is Germany's first sound film containing audible dialogue: *Das Land ohne Frauen* (*Bride Number 68*) (1929, dir. Carmine Gallone).

29 Some officially available versions should be treated cautiously though, as when (mostly commercial) restorations remove ground noise, add sound effects and even music, or otherwise alter the films in ways that make their historical analysis difficult.

30 One recent example is the retrospective "The Last Laugh: German Musical Comedies, 1930–32" at *Il cinema ritrovato*, Bologna 2022.

shooting and long takes were used to record whole scenes continuously rather than breaking them up into parts. While both multiple-camera shooting and static long takes in early sound film have often been regarded as a 'return to canned theatre,' as a deficiency that had yet to be overcome, Jacobs demonstrates how some directors, like Ford, used these techniques in intriguing ways. Long takes in particular "offered an attractive alternative to the stylistic infelicities of staging and framing for multiple cameras [...] freeing up the actor's movement and allowing for inventive staging in depth." In this perspective, the long take in early sound film does not so much appear as a 'step back' in the development of film aesthetics as an anticipation of later realist film styles.

Katharina Loew examines what she terms "montage shots" (superimpositions, split screen mattes, and prolonged lap dissolves) as stylistic devices that span from early cinema into the sound film era. As Loew states, these techniques attest "to a far greater consistency between silent and sound aesthetics than is usually acknowledged." Moreover, challenging the traditional conception of "montage sequences" as a device primarily used to compress time and space, she argues that in Weimar films like *Der brave Sünder* (dir. Fritz Kortner, 1931), scenes with montage shots were most frequently deployed to depict interior states and to "encourage viewers to forge conceptual connections between simultaneously presented images."

In the last chapter of this section, Donald Crafton demonstrates how pre-sound animation embraced "lexigraphic" devices such as "sound-suggestive hieroglyphs," *toponyms*, word and thought balloons, and "emoji *avant la lettre*," whereas the introduction of sound saw the demise of these techniques, with films now relying on what Crafton terms "melomania – an obsessive preoccupation with syncing screen action to a pre-recorded sound-track." Moreover, while pre-sound animation filmmakers primarily assimilated graphic traditions, such as comic strips, and adapted them to the specific affordances of film, *sound* animation capitalized on new kinds of "convergences" with live performance traditions, such as vaudeville. Crafton concludes that, at least with animation, the introduction of sound was indeed characterized by "disruptiveness," causing "a rapid and irrevocable change in styles, modes of production, and reception."

Section II, "From Theory to Practice," explores how film-theoretical conceptions about sound that were being formulated resonated in or were taken up by actual filmmaking. As is well known, the period was ripe with theoretical ideas surrounding the possible futures of sound films, but many of them did not immediately translate into film production. For instance, there was a five-year interim between the famous Russian

"Statement on Sound," which extended montage theory into the realm of sound by propagating its "contrapuntual" use,[31] and the release of the first sound film by one of its authors (*Dezertir*; dir. Vsevolod Pudovkin, 1933).[32] Some examples of more immediate transpositions are well known, such as the writings and films by René Clair in France, who advocated for a continuation of silent film aesthetics in the sound era through a sparing use of dialogue and sound.[33]

The chapters in this section present two very different case studies of lesser-known carryovers from theory into practice, demonstrating that theoretical propositions for sound film aesthetics would often come from unexpected directions: here, from the writings of a Russian emigré director in Hollywood and from an oral report by a Soviet film student and developer of sound film technology, who evokes the theoretical ideas of Russian Futurism.

In a case study of the early Hollywood talkie *Applause* (1929), Michael Slowik shows how director Rouben Mamoulian's formal resourcefulness was grounded in his theoretical conceptions of medium specificity. Drawing upon Mamoulian's published as well as unpublished writings preserved in the Library of Congress, Slowik argues that the director's devotion to stylization in the arts led him to conceptualize sound in terms of narratively expressive selection and organization rather than the mere recording of a pre-existing aural reality. Thus, as Slowik shows, *Applause* announced an unorthodox set of possibilities for film sound: overtly manipulated recorded voices, background music distorted to the point of grotesquerie, a "symphony" of noises to express a character's psychological state, and even on-location sounds selected and manipulated for narrative purposes.

Theoretical conceptions of sound as "meaningful expression" are also at the heart of Oksana Bulgakowa's chapter, which argues that several early Russian sound films drew on theoretical assumptions that are quite different from montage theory. She cites a report by Nikolai Anoshchenko held before the Association of the Workers of Revolutionary Cinematography, the transcription of which is preserved in the Russian State Archive. Invoking the expressive ideas of Russian Futurist poetry, Anoshchenko argues for a merger of music and voice in sound film and for phonosemantic practices in which words are chosen less for their actual meaning than for

31 Eisenstein, Pudovkin, and Alexandrov, "Statement on Sound."

32 It is perhaps even debatable how much this film can count as a realization of the ideas put forward in the manifesto. See Thompson, "Early Sound Counterpoint."

33 Clair, *Cinema Yesterday and Today*.

the expressive sonic values of certain consonants and vowels. Bulgakowa traces reflections of this concept in films and uncompleted film projects by Vsevolod Pudovkin and Sergei Eisenstein, as well as in four lesser-known productions by the state-run company Soyuzkino, all of which relied on the "phonetic expressivity of intonation" and on "musical vocalization." She interprets these efforts – derived from both theoretical thoughts and technological restraints – as a "homogenizing technique" that "levelled" all voices to a common standard and that would define Soviet sound films well into the 1940s.

Section III, "National Contexts," explores early sound film in Japan, France, and Germany, showing how the future of sound film aesthetics in each country was negotiated in written discourse *and* in practice, often in relation to silent film, theatre, and other electric sound media in the respective national contexts.

Johan Nordström traces efforts to develop a new aesthetics for Japanese sound film during the transitional period (which lasted as late as 1936), focusing on the production and reception of films from two major studios: Shoshiku and Nikkatsu. He concludes that "the Japanese film industry's extended transition facilitated aesthetic experimentation and a gradual shifting of representational styles and thematic concerns." Hybrid films, such as part-talkies or post-synchronized films (called "sound-version" films in Japan), were produced for a longer time compared with other major production countries and were debated widely by critics and filmmakers. Language and voices, in particular, were at the centre of practical experiments and critical debates, as more natural styles of elocution, using modern language and different dialects, were pivotal for a new form of "everyday realism" in modern melodrama, but they seemed at odds with the intended effect of historicity in the *jidaigeki* (period film) genre.

In a survey of some of the earliest sound films produced in France in 1930 and 1931, Martin Barnier demonstrates that even though contemporary critics such as René Clair and Georges Vial frequently dismissed French films from this period as a return to "canned theatre" and claimed a loss of silent film's visual eloquence, several of the films produced by the major production companies were in fact highly "experimental" in their deployment of sound and especially in their juxtapositions of sound and images. Fluent and elaborate camera movements, inventive use of direct and off-screen sound, as well as complex editing patterns characterize these films, especially at their beginnings. Moreover, several films self-consciously display the new medium of sound film, for instance in spoken opening credit sequences, and address its relation to other sound media of the time, such as the thriving record industry.

Jörg Schweinitz picks up the topic of early sound film's intermedial connections and outlines how both radio and early sound film figured prominently in the general enthusiasm for the period's new "audio culture" in Germany. As Schweinitz illustrates against the backdrop of contemporary discourses on the radio, the new mass medium was staged as a "visual sensation" in several German sound films of the early 1930s and thus "became part of the imagery of modernity." Often, radio functioned as a narrative device by providing "diegetic bridges" between narrative spaces, most notably between broadcasters and the modern "dispersed" audience as a (potentially) transnational listening community. Specifically, Schweinitz shows how the character of the radio reporter appears in several early sound films and how famous real-life reporter Alfred Braun, who plays himself in several productions, quickly became established as a "transmedial star" in Germany and as a "presenter of modern life, [...] who helped shape the imaginative world of urban modernity."

Section IV, "Speech and Language," is dedicated to issues of speech, language, and translation in early sound film. Irina Leimbacher traces the use of what she terms "individual embodied speech" in early Anglophone nonfiction films. While speaking subjects were common in newsreels of the early 1930s (if often 'faked' by reenactments and post-synchronization), the directors of the British documentary movement mostly strove to set their work apart from newsreels by relying on what they perceived as more artistic "treatments of reality," such as collages of noises and vocal fragments or modernist musical scores. Leimbacher discusses Arthur Elton and Edgar Anstey's *Housing Problems* (1935) as a singular example of a documentary from this period, in which sync sound voices are given "expository agency" by allowing working-class people to speak into the camera, addressing their own concerns and "lived experience." Interestingly, while the film was derided by some for a "lack of any aesthetic," Leimbacher points out that other contemporary reactions attest to "early recognition that embodied voices speaking from the screen *could* provide a profound experience." This opens up an interpretation of the film as a turning point in documentary's aesthetic, "eliciting other listening opportunities and affective engagement."

The next two chapters in this section explore the reception of multiple versions in the early sound era. As is well known, language barriers posed a severe problem for the international distribution of early talkies, a problem that producers initially aimed to overcome by shooting the same film in several versions (even if dubbing and subtitling also existed early on).[34] Primarily driven by economic imperatives, the practice had aesthetic implications, too. After all, what is "the" film in this case, how do the respective versions differ from each other aesthetically, and how did they affect different viewers from the respective language areas? The last two articles in this section tackle these questions from quite different perspectives and partly using different terminology.

Jessica Berry is interested in the reception of multilanguage versions (also known as multiple language versions or MLVs) in the multilingual context of Switzerland, where several versions of one film were shown, sometimes even in the same cities. As Berry explains, "local populations in Switzerland were receptive to foreign-language films due to the multilingual background of the country." But her study also shows that versions were not necessarily compared to each other, and that critics from each region regarded their version as 'the' film. In her examination of contemporary reviews of *Die Nacht gehört uns/La nuit est à nous* (1929), Berry also addresses the topic of noises in film, finding that Swiss critics from either language region were often (though not always) intrigued by the film's sounds of machines and engines rather than by its dialogues, which were often considered unnatural and dragging. Noises, on the other hand, were seen as "a quintessential reflection of the contemporary modern era" and as a move away from theatre-like productions.

Maria Adorno investigates the transcultural mediation processes involved in the production of European versions at the beginning of the sound era. She argues that film versions "tailored to the respective target 'mentalities'" involved not just linguistic but also – and even more importantly – cultural issues. As she illustrates with several film examples, omissions, mutual substitutions, and other forms of modification, as well as considerations regarding the specific actors' and actresses' voices and appearances played a role in adapting versions to diverging audience expectations. This leads Adorno to use the term "multiple versions" rather than "multilanguage versions" or "multiple language versions" and to construe the practice as a "mimetic technique," in which "each version has its own status of originality and its own impact in its respective target context." In a discussion of the aesthetic concept of mimesis, she argues for MVs as "simultaneously *one* transnational film and *many* national(ized) versions," comparable to a polyhedron, "whose unified structure is constituted by multiple interconnected faces."

34 See e.g., Wahl, *Multiple Language Versions Made in BABELsberg*; Vincendeau, "Hollywood Babel."

Finally, Section V, “Music and Noise,” explores the uses of music and background noises in early sound film, again focusing on specific national contexts and film genres. While the initial and primary fascination with sync sound rested on the human voice,[35] music and noises played a crucial role for new audiovisual aesthetics as well, even if that role was by no means clearly defined. Critics and filmmakers debated widely what the place of music and noise in sound film should and could be, while films tried out several of these options, opening up a whole range of different meanings ascribed to specific sounds.

While music in early sound film has most often been studied in relation to the musical genre and songs,[36] Daniel Wiegand explores how the general handling of nondiegetic music in early German sound films changed over a relatively short period of time. Examining a large corpus of more than sixty films, he finds that after an initial reluctance to use nondiegetic music (except in a limited set of genres and specific types of scenes), films from mid-1931 on began to score scenes more often and more liberally. This change was also reflected in the trade press, where several authors argued against nondiegetic music at first, before growing accustomed to the changes in film production.

The other two chapters in this section address the topic of noises in early sound films. Martin Holtz looks at the “sounds of war” in American and German war films of 1930, arguing that they constituted a “counter-discourse” to the notion of film sound as “progressive and utopian ‘electrical entertainment’” (the latter a term by Donald Crafton). By contrast, Holtz interprets the harrowing sounds of warfare in productions like *All Quiet on the Western Front* (dir. Lewis Milestone) and *Westfront 1918* (dir. G. W. Pabst) as “an encounter with technological modernity, only not in form of benevolent scientific advancement, but as traumatizing shock.” As Holtz demonstrates, often in these films, individuals are silenced by sound and subordinated to machines of sound production (such as the telephone) that they are forced to listen to.

Nadine Soraya Vafi explores urban noise as a vital part of the soundtrack in early Italian sound films. Based on the assumption that the “sounds of modern city life were characterized by simultaneousness and a fast-paced rhythm, and sound film, in particular, offered an apt sonic representation of these modern realities,” Vafi construes the metropolitan noises in the urban

35 This is an early instance of what Michel Chion has termed “vococentrism” in the cinema (Chion, *The Voice in Cinema*, 5–6).

36 See e.g., Altman, *The American Film Musical*; Spring, *Saying It with Songs*; O’Brien, *Movies, Songs, and Electric Sound*; Wedel, *Der deutsche Musikfilm*.

comedy *Gli uomini, che mascalzoni …* (1932, dir. Mario Camerini) as a reflection of Italian modernity through sound, especially of the socio-political changes emerging under fascism. Among other things, she discusses the expansion of highways and the showcasing of 'Italian strength and power' during large trade fairs. As she argues, rather than promoting fascism, the resulting soundscape should be seen as "part of a realist aesthetic that acknowledged working-class everyday life and concerns."

The articles by Holtz and Vafi highlight that the "world of noises"[37] opened up entire new, formerly excluded areas to filmic representation, almost as if to confirm Béla Balázs' utopian romanticism expressed in 1930: "What the sound film will now uncover is our acoustic environment […] everything that has something to say over and above human dialogue."[38] With the city and the war, two crucial spheres of experience became audible in a fresh manner for 1930s spectators, albeit in different ways, each of them complicating Balázs' anticipation of a harmonious interchange between humans and their environment through sound. While war noises arguably functioned as a way to cope with memory and trauma, the city noises in urban comedies brought a contemporary soundscape from just outside the cinema back into the auditorium, ironically one that was simultaneously shut out by the sound insulation of modern architecture.[39] And while war films staged the 'drowning out' of the individual through sound, city noise had the potential to situate the individual (especially from the working class) more firmly in lived urban space and thus to reaffirm it as a visible and audible part of modern society.

Work on this book goes back to 2021, when first ideas for it were put forward in a dense weekend of online presentations at the University of Zurich. I am very grateful to all the contributors, including the ones who are not present in the final volume, for sharing their expertise and their ideas, which – far beyond the individual papers – supported me greatly in the conception of this volume. I dearly hope that we will one day all meet in person!

I especially want to thank Jessica Berry and Nadine Soraya Vafi for their help in preparing the online event with me, as well as Margrit Tröhler, Fabienne Liptay, Yvonne Kummer, Denise Weber, Carla Gabrí, Elisabeth

37 This is the expression used by an anonymous critic in the German trade journal *Deutsche Filmzeitung*, January 10, 1930 (in the original: "Welt der Geräusche").

38 Balázs, *Early Film Theory*, 185.

39 See, for instance, Thompson, *The Soundscape of Modernity*.

Agethen, Philipp Doerler, Martin Weiss, Philipp Blum, and Simone Winkler from the Department of Film Studies in Zurich for additional support. I also thank the series editors of "Cinema and Technology," especially Katharina Loew and Santiago Hidalgo, for their support and the opportunity to publish this book with Amsterdam University Press, Maryse Elliot of Amsterdam University Press as well as Susie Trenka, Diliara Fruehauf, and Nico Uebersax for their great editorial support.

Bibliography

Abel, Richard, ed. *French Film Theory and Criticism, 1907–1939. Vol. II: 1929–1939.* Princeton, NJ: Princeton University Press, 1988.

Altman, Rick. *The American Film Musical.* Bloomington: Indiana University Press, 1987.

Altman, Rick. *Silent Film Sound.* New York: Columbia University Press, 2004.

Balázs, Béla. *Early Film Theory: "Visible Man" and "The Spirit of Film".* Ed. Erica Carter. New York: Berghahn, 2010.

Barnier, Martin. *En route vers le parlant: histoire d'une évolution technologique, économique et esthétique du cinéma (1926–1934).* Liège: Éditions du Céfal (Travaux & thèses, Cinéma), 2002.

Beck, Jay, and Tony Grajeda, eds. *Lowering the Boom: Critical Studies in Film Sound.* Urbana: University of Illinois Press, 2008.

Bordwell, David, Janet Staiger, and Kristin Thompson. *The Classical Hollywood Cinema: Film Style & Mode of Production to 1960.* New York: Columbia University Press, 1985.

Buhler, James, David Neumeyer, and Rob Deemer, eds. *Hearing the Movies: Music and Sound in Film History.* New York: Oxford University Press, 2010.

Chion, Michel. *The Voice in Cinema.* Translated by Claudia Gorbman. New York: Columbia University Press, 1999.

Christie, Ian. "Making Sense of Early Soviet Sound." *Inside the Film Factory: New Approaches to Russian and Soviet Cinema,* edited by Richard Taylor and Ian Christie, 176–92. London/New York: Routledge, 1991.

Clair, René. *Cinema Yesterday and Today,* edited by R. C. Dale. New York: Dover Publication, 1972.

Cooke, Mervyn, and Fiona Ford, eds. *The Cambridge Companion to Film Music.* Cambridge: Cambridge University Press, 2016.

Crafton, Donald. *The Talkies: American Cinema's Transition to Sound 1926–1931.* Berkeley: University of California Press, 1997.

Davidson, John E., and Theodore F. Rippey, eds. "Early Sound Cinema in the Late Weimar Republic." *Colloquia Germanica* 44, no. 3, (2011): 233–36.

Eisenstein, Sergei, Vesvolod Pudovkin, and Grigori Alexandov. "Statement on Sound." Sergei Eisenstein. *Selected Works vol. 1: Writings 1922–34*, edited and translated by Richard Taylor, 113–14. London: I. B. Tauris, 2010.

Elsaesser, Thomas. "Going 'Live': Body and Voice in Some Early German Sound Film." In *Le son en perspective: Nouvelles recherches / New Perspectives in Sound Studies*, edited by Dominique Nasta and Didier Huvelle, 155–68. Brussels: P.I.E.-Peter Lang, 2004.

Fleeger, Jennifer. *Sounding American: Hollywood, Opera, and Jazz*. New York: Oxford University Press, 2014.

Gunning, Tom. "'From the Bottom of the Sea': Early Film at the Oberhausen Festival." In *Early Cinema Today: The Art of Programming and Live Performance*, edited by Martin Loiperdinger, 37–41. New Barnet: John Libbey, 2011.

Helmers, Maike. "The Transition from the Silent into the Sound Era: The Innovative Use of Sound in Pabst's *Westfront 1918*." *Music, Sound, and the Moving Image* 12, no. 2 (2018): 121–39.

Hanson, Helen. *Hollywood Soundscapes: Film Sound Style, Craft and Production in the Classical Era*. London: Bloomsbury, 2017.

Huhtamo, Erkki and Jussi Parikka, eds. *Media Archaeology: Approaches, Applications, and Implications*. Berkely/Los Angeles/London: University of California Press, 2011.

Iwamoto, Kenji. "Sound in the Early Japanese Talkies." In *Reframing Japanese Cinema: Authorship, Genre, History*, edited by Arthur Nolletti, Jr. and David Desser, 312–27. Bloomington: Indiana University Press, 1992.

Jacobs, Lea. *Film Rhythm After Sound: Technology, Music, and Performance*. Berkeley: University of California Press. 2015.

Jacobs, Lea. "The Innovation of Re-Recording in the Hollywood Studios." *Film History* 24, no. 1 (2012): 5–34.

Kaes, Anton, Nicholas Baer, and Michael Cowan, eds. *The Promise of Cinema: German Film Theory 1907–1933*. Berkeley: University of California Press, 2016.

Kaganovsky, Lilya. *The Voice of Technology: Soviet Cinema's Transition to Sound 1928–1935*. Bloomington, Indiana: Indiana University Press, 2018.

Lewinsky, Mariann. "The Best Years of Film History: A Hundred Years Ago." In *Early Cinema Today: The Art of Programming and Live Performance*, edited by Martin Loiperdinger, 25–35. New Barnet: John Libbey, 2011.

Lewis, Hannah. *French Musical Culture and the Coming of Sound Cinema*. New York: Oxford University Press, 2018.

Moure, José. "Du silence au cinéma." In *Voix et média*, edited by Marie Thonon, 24–38. Paris: L'Harmattan, 1999.

Müller, Corinna. *Vom Stummfilm zum Tonfilm*. München: Wilhelm Fink, 2003.

Nasta, Dominique, and Didier Huvelle, eds. *Le son en perspective: Nouvelles recherches/New Perspectives in Sound Studies*. Brussels: P.I.E.-Peter Lang, 2004.

Nordström, Johan. "Between Silence and Sound: The Liminal Space of the Japanese 'Sound Version.'" In *Routledge Handbook of Japanese Cinema*, edited by Joanne Bernardi & Shota T. Ogawa, 213–30. London/New York: Routledge, 2021.

Porter, Laraine. "Okay for Sound? The Reception of the Talkies in Britain 1928–1932." *Journal of British Cinema and Television* 17, no. 2 (2019): 212–32.

O'Brien, Charles. *Cinema's Conversion to Sound: Technology and Film Style in France and the U.S*. Bloomington and Indianapolis: Indiana University Press, 2005.

O'Brien, Charles. *Movies, Songs, and Electric Sound: Transatlantic Trends*. Bloomington: Indiana University Press, 2019.

O'Rawe, Des. "The Great Secret: Silence, Cinema and Modernism." *Screen* 47, no. 4 (2006): 395–405.

Raine, Michael, and Johan Nordström. *The Culture of the Sound Image in Prewar Japan*. Amsterdam: Amsterdam University Press. 2020.

Salt, Barry. *Film Style & Technology: History and Analysis*. London: Starword, 1983.

Slowik, Michael. *After the Silents: Hollywood Film Music in the Early Sound Era, 1926–1934*. New York: Columbia University Press, 2014.

Spadoni, Robert. *Uncanny Bodies: The Coming of Sound Film and the Origins of the Horror Genre*. Berkeley: University of California Press, 2007.

Spring, Katherine. *Saying It with Songs: Popular Music and the Coming of Sound to Hollywood Cinema*. New York: Oxford University Press, 2013.

Street, Sarah, and Joshua Yumibe. *Chromatic Modernity: Color, Cinema, and Media of the 1920s*. New York: Columbia University Press, 2019.

Szczepanik, Petr. "Sonic Imagination: or, Film Sound as a Discursive Construct in Czech Culture of the Transitional Period." In *Lowering the Boom: Critical Studies in Film Sound*, edited by Jay Beck and Tony Grajeda, 87–104. Urbana: University of Illinois Press, 2008.

Thompson, Kristin. "Early Sound Counterpoint." *Yale French Studies*, no. 60 (1980): 115–40.

Thorburn, David, and Henry Jenkins. *Rethinking Media Change: The Aesthetics of Transition*. Cambridge: MIT Press. 2004.

Vincendeau, Ginette. "Hollywood Babel: The Multiple Language Version." *Screen* 29, no. 2 (1988): 24–39.

Wahl, Chris. *Multiple Language Versions Made in BABELsberg: Ufa's International Strategy, 1929–1939*. Amsterdam: Amsterdam University Press; Eye Filmmuseum, 2016.

Weis, Elisabeth. *The Silent Scream: Alfred Hitchcock's Sound Track*. Rutherford: Fairleigh Dickinson University Press, 1982.

Weis, Elisabeth, and John Belton, eds. *Film Sound: Theory and Practice*. New York: Columbia University Press, 1985.

Wedel, Michael. *Der deutsche Musikfilm: Archäologie eines Genres 1924–1945*. München: edition text + kritik, 2007.

Wedel, Michael. *Pictorial Affects, Senses of Rupture: On the Poetics and Culture of Popular German Cinema, 1910–1930*. Berlin: De Gruyter, 2019.

Wiegand, Daniel. "The Delightful Paradox: Towards an Aesthetics of Silence in Early Sound Film." In *Silence, Sounds, Music: Acoustic Dimensions of Immersion*, edited by Peter Niedermüller, Florian Freitag and Laura Mücke. New York: Routledge, forthcoming.

Wiegand, Daniel. "'Eine Entdeckungsfahrt in die Welt der Geräusche': Zur historischen Rezeption des frühen Tonfilms *Die Nacht gehört uns*." *Montage AV* 30, no. 2 (2021): 59–76.

Wiegand, Daniel. "'Islands of Sound in the Silent Flow of Film': German Part-Talkies around 1930 as a Hybrid Medium." *Historical Journal of Film, Radio, and Television, Historical Journal of Film, Radio, and Television* 24, no. 3 (2022): 427–50. https://doi.org/10.1080/01439685.2021.1976916.

Wiegand, Daniel. "Listening to Faint Sounds and Silence: Cinema's Transition to Sound and the Emergence of a New Auditory Sensitivity." In *Retuning the Screen: Sound Methods and the Aural Dimension of Film and Media History,* edited by Simone Dotto, 85–94. Milan: Mimesis International, 2022.

Wood, Nancy. "Towards a Semiotics of the Transition to Sound: Spatial and Temporal Codes." *Screen* 25, no. 3 (May–June 1984): 16–25.

Wurtzler, Steve. *Electric Sounds: Technological Change and the Rise of Corporate Mass Media.* New York: Columbia University Press, 2007.

About the Author

Daniel Wiegand is Assistant Professor of Film Studies at University of Zurich. He is the author of *Gebannte Bewegung: Tableaux vivants und früher Film in der Kultur der Moderne* as well as co-editor of *Film Bild Kunst: Visuelle Ästhetik im vorklassischen Stummfilm* (edited with Jörg Schweinitz, 2016). Recent articles include "'Islands of Sound in the Silent Flow of Film': German Part-Talkies around 1930 as a Hybrid Medium," in *Historical Journal of Film, Radio, and Television* (2022).

I

From Silence to Sound

2. Dialogue Scenes in the Period of Multiple-Camera Shooting

The Example of *Arrowsmith*

Lea Jacobs

Abstract: Multiple-camera shooting was the primary method of shooting synchronized sound in the American cinema from the earliest Vitaphone shorts to the end of 1931. This chapter analyses the reasons for its use and the difficulties and advantages it posed for staging, shooting and cutting dialogue scenes in this period. The multiple-camera technique is contrasted with two alternative methods: the typically fast-cut conversation scenes of the late silent cinema and sound filming with a single camera in a single take. A detailed analysis of John Ford's *Arrowsmith* (1931) considers the implications of the long-take technique for staging and performance.

Keywords: speech in early sound film, camerawork, John Ford, long takes, acting

Multiple-camera shooting was the primary method of shooting synchronized sound from the earliest Vitaphone shorts to the end of 1931. Several technical problems contributed to the practice. In order to avoid recording camera noise, the camera was initially sequestered in a booth or "ice box" (Fig. 2.1) and somewhat later, encased in a heavy blimp.[1] Moving the camera around the set to create new setups thus became difficult and time-consuming. Another motivation for the practice was the pronounced preference for uninterrupted sound takes in the early sound cinema. Editing sound was difficult prior to the invention of edge numbering and other devices for maintaining

1 Bordwell, Staiger, and Thompson: *The Classical Hollywood Cinema*, 298–308.

Wiegand, D. (ed.), *Aesthetics of Early Sound Film: Media Change around 1930*.
Taylor & Francis Group, 2023
DOI 10.5117/9789463727372_CH02

Figure 2.1: Camera encased in booth on set (production still courtesy of the Wisconsin Center for Film and Theater Research).

synchronization.[2] Moreover, the relevant contemporary models for recording all worked by live, real-time sound capture. In the 1920s and 1930s radio was most often broadcast live. While engineers could mix multiple channels after mixing consoles came into use in the middle 1920s, this process was still intrinsically bound to the real time and space of performance.[3] Similarly phonograph discs were directly recorded in a single session.[4] In the USA, it was not until after World War II that the widespread use of magnetic tape provided a medium for mixing sound prior to broadcast and editing music tracks in the production of records.[5] Film was thus effectively the first medium that afforded extended opportunities for the recombination of sounds after the recording stage, but, as I have argued, this was not institutionalized as a practice until after RCA's innovation of the push-pull track in 1935.[6] Prior to this time, rerecording lowered the quality of the resulting sound track beyond

2 Salt, *Film Style & Technology*, 234–35.
3 VanCour, *Making Radio*, 91–5.
4 Millard, *America on Record*, 286.
5 Millard, *America on Record*, 199–201 and 289–308; Copeland, *Sound Recordings*, 26–30.
6 Jacobs, "The Innovation of Re-recording," 29.

Figure 2.2: Sound engineer in mixing booth on set (production still courtesy of the Wisconsin Center for Film and Theater Research).

what was considered acceptable, especially for the rendition of dialogue. Instead, mixing occurred in real time on the sound stage as it did in radio (Fig. 2.2 shows the sound engineer in the booth on a film set).

Figure 2.3 shows a present-day multiple-camera setup used for streaming or shooting a live television broadcast. In the early sound period, the cameras would have been enclosed in some way, and one camera would have filmed the sound while others would have simultaneously photographed the action with different framings and slight variations in the angle of view. The editor would then have cut the picture track, matching it to the uninterrupted sound record. During multiple-camera shooting, the cameras tended to be stationary, although footage shot 'wild' (without sound) was frequently inserted to introduce camera and figure movement before, after, or during a pause in a sync-sound conversation. It was also possible to pan or tilt cameras within a booth, or to move blimped cameras by putting them on a dolly. For example, *A House Divided* (1931, dir. William Wyler) used dolly shots to great effect in several scenes in which the father (Walter Huston) and son (Kent Douglass) argue while walking. But most commonly, not only were the cameras fixed in position, but also the movements of the actors were restricted during conversation scenes since they had to remain in view of all cameras simultaneously.

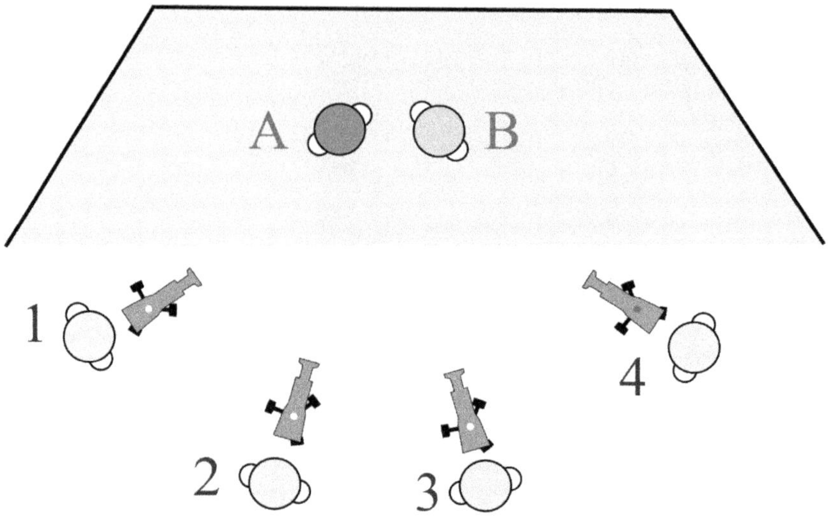

Figure 2.3: Multiple-camera setup for live television broadcast.

Multiple-camera shooting also diminished the sense of depth that could be achieved in the shot-reverse-shot configuration. It was hard for the cameramen to get in close to the actors for singles or over-the-shoulder shots since it was necessary to keep each camera out of the angle of view of the others. This meant editors did not have a nice variation of angle as they cut between interlocutors. David Bordwell has demonstrated the effects of multiple-camera shooting in a scene from the Warner Brothers film *The Lights of New York* (1928, dir. Bryan Foy).[7] Figure 2.4 shows the long-shot framing. This is the only framing which allows for some figure movement as actors enter and exit from the rear door. The medium-long-shot framing in Figure 2.5 shows the three interlocutors in conversation. Figures 2.6 and 2.7 show how the space of the three-shot was broken up into two laterally adjacent closer views, the multiple-camera equivalent of shot-reverse-shot. Note that the space appears flat in the cuts between interlocutors – we seem to move from side to side rather than taking up the position of first one speaker and then the other. During the conversion to sound, some directors such as John Ford, Edmund Goulding, John M. Stahl and George Cukor experimented with filming dialogue scenes from a single camera setup in a long take. By this means they avoided the awkward framings and side-by-side cutting between adjacent views that we see in *Lights of New*

7 Bordwell, Staiger, and Thompson: *The Classical Hollywood Cinema* 305–6.

Figure 2.4: *Lights of New York* shot 5, establishing shot.

Figure 2.5: *Lights of New York* shot 6, three-shot.

Figure 2.6: *Lights of New York* shot 7, two-shot.

Figure 2.7: *Lights of New York* shot 8, single.

York. Of this group two, Ford and Stahl, experimented extensively with a fixed camera position as well.

Take, for example, a shot of almost three minutes from *Seed* (1931, dir. Stahl). Lois Wilson, playing Peggy Carter, a mother of five, stands in back of her stove making soup while she confronts a rival for her husband's affections. The rival, Mildred, played by Genevieve Tobin, paces from foreground to background, hypocritically lauding Peggy's housekeeping skills and singing the praises of Peggy's husband (Fig. 2.8 – Fig. 2.11). Long takes such as Stahl employed here offered an attractive alternative to the stylistic infelicities of staging and framing for multiple cameras. The actor could move with greater freedom and shots could be composed in greater depth. Moreover, as the case of *Seed* indicates, the single, fixed take elevated the importance of small gestures by the actors. As she speaks or listens, Carter lifts the pot lid and stirs the soup, affirming her wifely and maternal role

Figure 2.8: *Seed* long take near beginning.

Figure 2.9: *Seed* long take after Tobin walks into background.

Figure 2.10: *Seed* long take after Tobin returns to foreground.

Figure 2.11: *Seed* long take near end.

(Fig. 2.10). Tobin responds by moving forward, lowering her veil and showing off her fur purse (Fig. 2.11).

For Ford, as for Stahl, the static long take provided a way of freeing up the actor's movement and allowing for inventive staging in depth. Although his films demonstrate an interest in composition in depth from the beginning of his career in the silent period, Ford was not, on the whole, a long-take director. He began making films at a point in the 1910s when the American film industry was in transition from a long-take or tableau style of filmmaking to a shot-based mode of scene construction.[8] In contrast with a previous generation of directors such as his mentors D. W. Griffith and his brother

8 Salt, *Film Style & Technology*, 99–109, 149–54; Bordwell, Staiger, and Thompson: *The Classical Hollywood Cinema*, 155–230.

Francis Ford, the younger man readily embraced the new conventions of cutting around within a scene. Even his earliest feature, *Straight Shooting* (1917), while employing some longer takes as was typical of the period, employs both fast Griffith-style cross-cutting during last-minute rescues and, more unusually for 1917, a real facility with scene dissection. By the 1920s, his films have average shot lengths (ASL) of between 5 and 6 seconds, a rate in accord with Hollywood norms for this decade.[9] The ASL for *The Iron Horse* (1924), for example, is 5.5 seconds and for *3 Bad Men* (1926) is 5 seconds. His evolution in the direction of longer takes was initiated with *4 Sons* (1928), shot silent with an added music-and-effects track, and then solidified with his early sync-sound films. *4 Sons* has an ASL of 7.6 seconds, an increase in average shot length of about 40 per cent over the silent films cited above. This increase is pretty clearly a result of his experimentation with camera movement. In contrast, the silent *Hangman's House* (1928), made after *4 Sons*, has an ASL of 5.3. The return to faster cutting rates is partly due to the very-fast-cut horse race in the second act and partly to the reduction in the number and extent of camera movements. Although *4 Sons* does have some pronounced static long takes as well as elaborate camera movements, it is not until the early sound films that Ford seems to experiment extensively with static long takes in conversation scenes.

It is important to recognize the novelty of this general approach to shooting conversation scenes for directors and actors used to the conventions of silent cinema. In the late 1920s dialogue scenes without intertitles, though possible, and of course championed by Carl Mayer, F. W. Murnau and others, were rare. For most filmmakers most of the time, staging and shooting dialogue scenes involved anticipating the placement of intertitles by editors and title writers. Ford told Peter Bogdanovich: "In silent pictures we always had the actors speak the lines they were supposed to say anyway because there were too many lip readers in the audience."[10] Writing in 1919, the Danish director Urban Gad explained the way dialogue titles were inserted into scenes:

> Of course, the dialogue [title] must be so conceived as to match precisely with the acting [...] it must also be introduced precisely when it appears from the actors' expressions or movements that they are speaking just those words. Before a dialogue title is introduced, the actors have to be examined carefully over quite a long stretch of the film so as to locate

9 Bordwell, Staiger, and Thompson: *The Classical Hollywood Cinema*, 304; Salt, "Barry Salt's Database."

10 Bogdanovich, *John Ford*, 50.

> the point at which glances and movements of the mouth and hands are appropriate. This point also has the advantage that the listener in the film will also have the right facial expression, since the director will have envisaged similar dialogue during shooting.[11]

That is, the editor sought to place the intertitle so that it "matched" the line spoken by the actors. By the same token, the director and actors had to anticipate such a cut. Conversation scenes unfolded with the actors heightening and playing to the lines likely to be titled.

Consider this interchange between Charles Farrell as Chico and Janet Gaynor as Diane in *7th Heaven* (1927, dir. Frank Borzage). He has unexpectedly brought home a wedding dress for her, and her initial reaction as she fingers the dress and he proudly looks on is a bravura piece of silent film acting, largely without titles. As the scene proceeds, however, titles become necessary especially since the conversation revolves around a speech act. As she turns sad and sinks to the floor, they begin to talk, their conversation built up of fragments, cutting between Chico, Diane and the corresponding intertitles (Fig. 2.12 – Fig. 2.22). Farrell advances towards Gaynor, preparing to ask a question. Cut to the title: "Don't you want to marry me?" Cut to Gaynor, who nods slightly as she slowly turns to face him. Cut to the title: "But you never said – you love me." Cut back to Gaynor as she turns away. Cut to Farrell fumbling with his pipe in puzzlement. Cut to Gaynor as she prepares to speak, still facing away from Farrell. Cut to title: "Couldn't you say it – just once?" The scene proceeds with a stately rhythm which belies how quickly it is cut. This 33 second segment contains 8 shots including titles, yielding an average shot length of 4 seconds, a cutting rate just faster than the norm of 5 seconds. Indeed, I suspect the need to edit dialogue scenes around the intertitles in the silent period helped to standardize 5 seconds as the norm.

A scene from *4 Sons* illustrates Ford working around the alternation of picture and title, which was at the core of dialogue scenes in the American silent cinema. In the scene, Colonel Von Stomm (Earle Fox) confronts Mother Bernle (Margaret Mann) and her youngest son Andreas (George Meeker) about the fact that her emigrant son Josef is fighting on the American side in World War I. The scene contains a relatively long take of 24 seconds near the beginning and a very long take of 50 seconds at the end, but most of the scene is staged and acted in short takes, piecemeal fashion, in relation to the most important lines of dialogue. Following his entrance, Earle Fox as

11 Gad, *Der Film*, 245–46 (my translation).

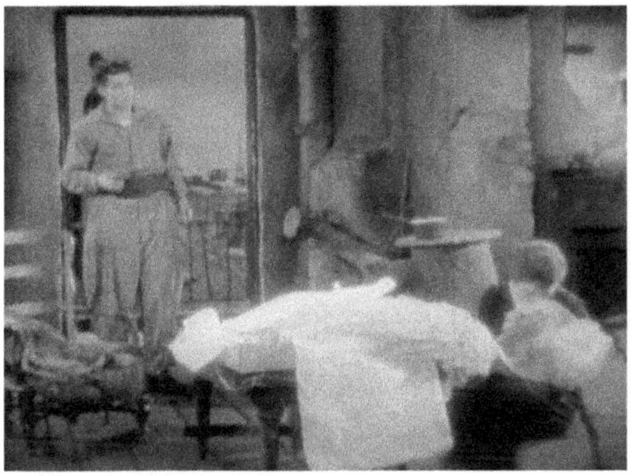

Figure 2.12: *7th Heaven* shot 1a.

Figure 2.13: *7th Heaven* shot 1b.

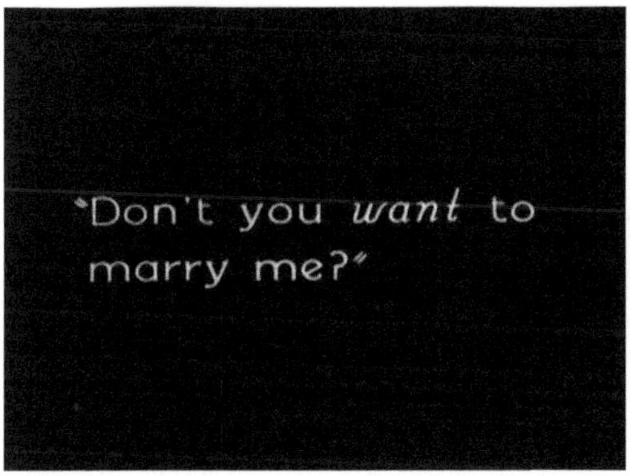

Figure 2.14: *7th Heaven* shot 2.

Figure 2.15: *7th Heaven* shot 3a.

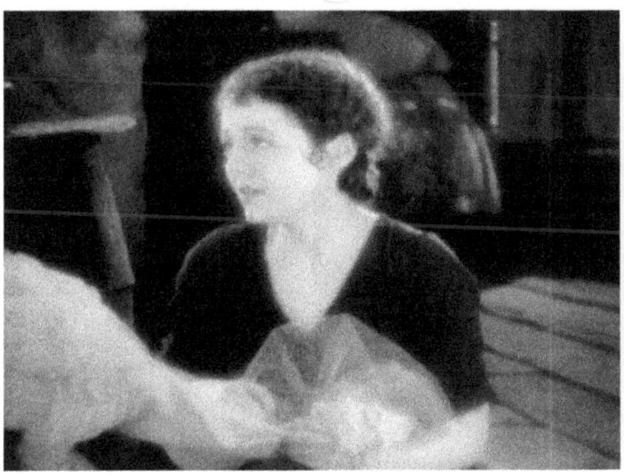

Figure 2.16: *7th Heaven* shot 3b.

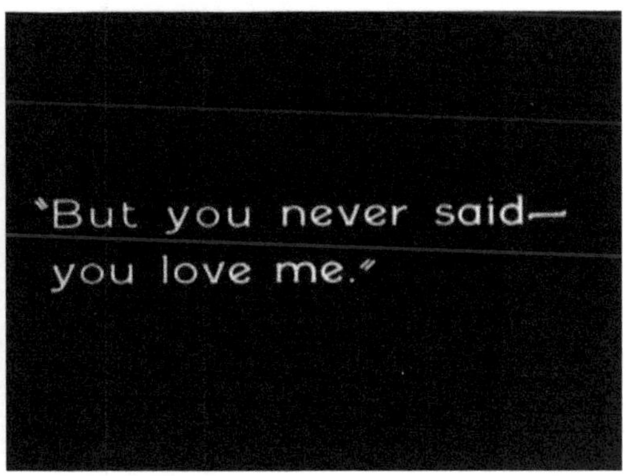

Figure 2.17: *7th Heaven* shot 4.

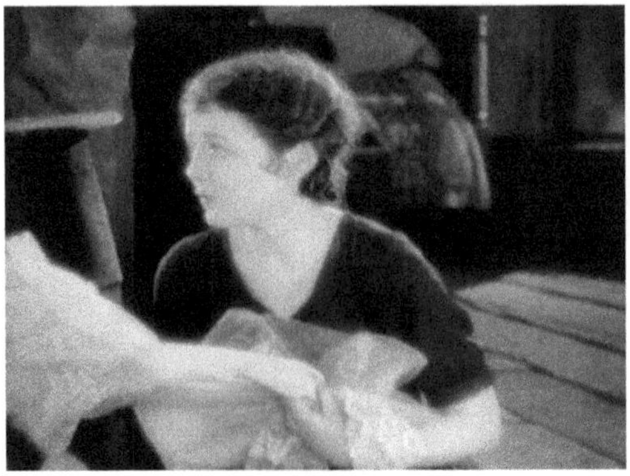

Figure 2.18: *7th Heaven* shot 5a.

Figure 2.19: *7th Heaven* shot 5b.

Figure 2.20: *7th Heaven* shot 6.

Figure 2.21: *7th Heaven* shot 7.

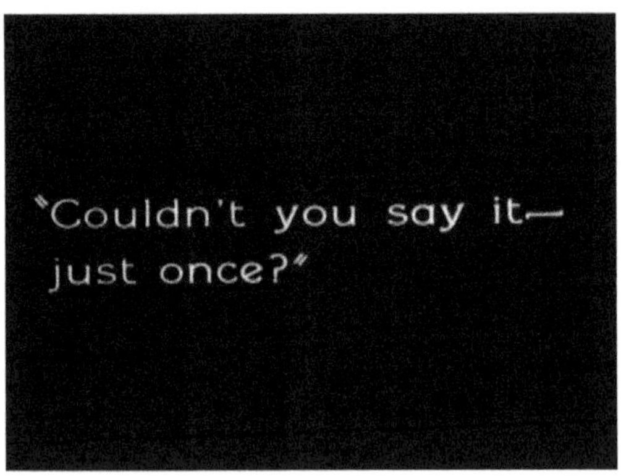

Figure 2.22: *7th Heaven* shot 8.

the commander takes a leisurely 24 seconds to build up to the accusation addressed to Mother Bernle in long shot: "You have a son in America. Give me his letters!" After the title, cut to a one-second reprise of the long shot. Cut to the mother's reply: "I have not heard from him in many months." Cut to a six-second closer view of the officer, his swagger stick reaching out to the mother in front of him, followed by the title: "The better for you! He is a traitor!" and return to the closer framing of Von Stomm for one second. Cut to a close-up of the mother's reaction for 10 seconds.

As the Colonel continues to insult Mother Bernle, cut away to Andreas who gestures to two framed government tributes on the wall and speaks to a sympathetic younger officer, an eight-second shot. Cut to the title: "He forgets my brothers Franz and Johann. They died fighting for the Fatherland." Cut back to Andreas and his interlocutor for 4 seconds.

In the scene's last shot, the fast-cut middle section gives way to the mother's prolonged reaction to the Colonel's plan to draft Andreas as recompense for Josef's enlistment in the US army. At the end of the second-to-last shot the Colonel gestures at Andreas with his swagger stick followed by the title: "Report at the barracks tonight! You will take your brother's place!" Cut back to a longer framing of the same gesture. As the Colonel snaps down the swagger stick, the mother comes from off screen to stand between him and her son. She comes forward and gestures to the government tributes previously indicated. She then gives way to expressions of grief as the men depart. The dialogue titles which are crucial to our comprehension of this scene are thus packed into its central portion, but the editing builds to a long-take finish which focuses on the mother's responses to the Colonel enacted without interruption – no intertitles, no close-ups.

The introduction of sync-sound obviated the need for the manipulation of title placement of the sort that Ford resorted to in *4 Sons*. *Arrowsmith*, shot for Goldwyn in the summer and fall of 1931, exemplifies a range of treatments of dialogue scenes. Although Ford employed the multiple-camera technique in many conversations, he also experimented extensively with staging and timing dialogue scenes in single static takes. His long-take strategy led to an average shot length of 12.5 seconds for the film as a whole, significantly slower than *4 Sons*, although *Arrowsmith* contains many fewer shots with camera movement.

Consider the scene in which Martin Arrowsmith (Ronald Colman), a country doctor, is inspired to return to his early interest in medical research by an outbreak of blackleg which is killing off the cattle of local farmers. His attempt to improve on the ineffectual serum being administered by the state veterinarian will ultimately be successful and garner him national recognition and an appointment at a major research facility. Colman is first shown in close-up at the microscope. He looks up and calls for his wife Leora (Helen Hayes). A cut to long shot reveals that the kitchen has become a makeshift laboratory (Fig. 2.23). The tubing that stretches from the stove in the foreground to the table in the background helps to transform the domestic space and establish that the pots are being used for scientific purposes. Leora enters from off screen left, and their conversation unfolds in a long take of just over one minute. The shot may be divided into two parts: the first is the scientist's monologue punctuated by Leora's repeated phrase of assent "Yes, Martin." The second part is Leora's attempt to engage Martin in conversation on the subject of dinner.

After Hayes's entrance, Colman comes forward on the line: "Now *I'm* ready to go. We'll try mine [my serum] in the morning," dropping a slide into the boiling kettle in the front left corner for sterilization (Fig. 2.24). He then moves to the background, fetches a towel from behind a half-opened door (Fig. 2.25), then comes forward again, using the towel to take hold of the hot flask in which he cooks his serum. He talks about injecting the medicine into a farmer's cattle (Fig. 2.26).

Colman returns to the background once again, dodging around the tubing, to set the serum on the table. He moves farther back into the area in front of the microscope, leans forward, and taps a cigarette on the table creating an aperture framing within the tubing (Fig. 2.27). In this position, he begins to explain his plan for administering different dosages to the cows. Then he comes forward, stopping to light a match just prior to the most important phrase: "And then we can see which gets well quickest and which dies quickest" (Fig. 2.28). This establishes the necessity of a control group in medical trials, an important point for the subsequent development of the plot. Then, exhaling a cloud of smoke, he returns to the table in the background.

Thus, in the first half of the shot Colman synchronizes his words and his gestures as he moves around the set. From the point of view of the actor, one of the great advantages of having a set designed for a single as opposed to multiple cameras is the deep playing space. Colman moves from the table to the stove and back again three times, each time coming forward to deliver an emphatic line or phrase, all the while manipulating props – the slide, the flask, the cigarette – and dodging the tubing. Thus, he demonstrates his control over the kitchen, previously Leora's domain, and his practiced use of the instruments of his trade. While his movements may seem ordinary, incidental, they are in fact carefully timed in relation to his lines, and serve important dramatic functions.

Figure 2.23: *Arrowsmith* near beginning of long take.

Figure 2.24: *Arrowsmith* long take, "I'll try mine in the morning."

Figure 2.25: *Arrowsmith* long take, Colman walks to background.

Figure 2.26: *Arrowsmith* long take, "I'm going to take it over to Henry's and shoot all his cattle full."

Figure 2.27: *Arrowsmith* long take, "I'm going to try a different strength does on every cow."

Figure 2.28: *Arrowsmith* long take, "And then we can see which gets well quickest and which dies quickest."

Figure 2.29: *Arrowsmith* long take, "Are you hungry?"

Figure 2.30: *Arrowsmith* long take, four-second pause.

Figure 2.31: *Arrowsmith* long take, "How can I cook dinner when you're using the whole stove to cook your old cow medicine!"

In the second part of the shot, Martin finally stops talking and settles down to work at the table. The timing reaches comic proportions as Hayes attempts to bring up the topic of dinner in the face of her husband's abstracted indifference. Standing in the foreground with back to camera (Fig. 2.29), she speaks: "Darling, are you hungry?" *Two-second pause*. "Darling are you hungry?" [*Louder*]. There is a four-second pause as Colman takes his time responding. He works head down, then hems and haws before finally looking up from his work, pencil in hand (Fig. 2.30). The sound of the pencil dropping on the table emphasizes his response: "Hungry. Yes, by gosh, I am famished." He resumes his pencil and his work, "Let's have some dinner." Hayes turns to camera revealing her face (Fig. 2.31). Exasperated, she asks: "How can I cook dinner when you are using the whole stove to cook your old cow medicine?" His slow and distracted responses eventually drive her out of the kitchen. She exits left, followed by the sound of a door slamming off.

Hayes questions, waits, questions. Colman hesitates, lifts his head, hesitates with pencil in hand, until he finally confesses to being hungry. These subtle calculations of the actors' craft lend a sense of spontaneity and freshness to their interactions. In the highly edited form that was the norm in silent cinema, the timing of the conversation, the pauses and the subtle interplay of gestures would have been largely controlled by the editor and the director in post-production. This is not to say that silent film acting was not skilled, but rather that it occurred piecemeal, and that the timing across the arc of a conversation was not under the actor's control as it would have been on the stage. The long-take style of the early sound period restored some of that control over timing.

In a later scene from *Arrowsmith*, Ford used a modified form of the multiple-camera technique in much the same way as the long take in the kitchen scene – to allow the performance to unfold in continuity. The scene involves Arrowsmith and his colleague Sondelius (Richard Bennett) who have travelled to the West Indies where there is an outbreak of bubonic plague. Arrowsmith has been charged with conducting a trial of his new serum, inoculating only half of the population of the island and administering a placebo to the other half as a control.[12] Sondelius has refused to take the serum. This decision is not motivated in the film but the 1925 novel on which it is based, by Sinclair Lewis, makes explicit that Sondelius opposes the use of a control group and will not take the serum himself until everyone can get it. Sondelius contracts the plague and dies in the makeshift hospital that he has constructed, tended by Arrowsmith and surrounded by patients and assistants who mourn his passing.

The conventional way to treat a tender death scene like this would be to use shot-reverse-shot. Even if Arrowsmith does not say very much, you would expect to see his face in close-up and see him react to his dying friend's words. And as it was typically used in this period, multiple-camera shooting

12 In both the film and Sinclair Lewis's original novel the situation in the West Indies is explicitly racialized, although the film seems more aware of the potentially offensive aspects of this. In both film and novel, the white governor of the island refuses to allow the experiment to proceed on the largely white population. In the novel, a white plantation owner then invites the scientists to conduct their experiment on the Black workers on his plantation. In the film, Dr. Marchand (Clarence Brooks), a Black physician, approaches Arrowsmith and Sondelius and suggests that they conduct the experiment on "his people," the largely Black population of the island of Caribe. In both novel and film, Arrowsmith finally breaks down and administers the serum to all who want it, abandoning experimental protocol.

was designed precisely to preserve the possibility of this kind of editing. But although this scene employs the multiple-camera technique, there is no shot-reverse-shot and no close-up of Colman. The death of Sondelius was shot simultaneously from two basic camera positions both centred on Bennett: a frontal long shot (Fig. 2.32, setup 1) and a medium shot taken with a long lens from the right (Fig. 2.33, setup 2).[13] An additional close-up of Bennett (setup 3) was taken singly. The editing created three long takes out of seven total shots as follows:

Shot 1	Long shot	Setup 1	10.23 seconds
Shot 2	**Medium shot**	**Setup 2**	**52.29 seconds**
Shot 3	Long shot	Setup 1	11.75 seconds
Shot 4	Medium shot	Setup 2	13.54 seconds
Shot 5	**Close-up**	**Setup 3**	**29.75 seconds**
Shot 6	Medium Shot	Setup 2	4.79 seconds
Shot 7	**Long shot**	**Setup 1**	**41.33 seconds**

The long shot framing (setup 1) which recurs in shots 1, 3 and 7 is held for over 41 seconds at the end of the scene. Shot 7 shows Arrowsmith's actions after the death as he covers the body and allows himself a brief expression of grief (Fig. 2.34) before returning to his labours. Shots 2 and 5, the other two long takes, concentrate on Bennett. In the shooting script, Bennett's lines were shorter and less colloquial than in the film. I surmise that Ford or Bennett (an experienced stage actor) expanded the length of Sondelius's speeches in performance, making them more vivid and colloquial. Shot 2 (Fig. 2.33), the longest shot in the sequence, is largely comprised of the dying man's reminiscence of his homeland and follows on from Bennett's line in shot 1, "Martin, you ever think of the jokes God plays?"

> The best one is on the tropics. He made them so beautiful and so rich and then he give them the plague ... Oh I ... I am quite peaceful. Martin did you ever see Stockholm? I would like yust once more to see the Strandvägen at dawn with the young snow, falling, white ... And stagger home through it with one more last good drunk. Yeah, eh ...[14]

13 *Arrowsmith* Production Reports, Folder 26, Box 4. I consulted the Daily Production Report and the Stage Log for September 24, 1931, which indicate that scene number 177 was shot with two cameras, one long shot and one medium.

14 The script is written in full and grammatically correct sentences. The actor added the Swedish pronunciation and also introduced more colloquial phrasing. I attempt to capture this in my transcription of the film's dialogue.

The unique closer framing in shot 5 helps to drive home Bennett's dying plea to his friend to abandon the controlled experiment and give all the occupants of Caribe a chance of being cured:

> You try to save all these poor devils [*Rain sound diminishes over speech.*] Save all of them! Let science ... let experiments go. Huh, I never know before people could hurt me so much. [*Thunder. Rain sound increases.*] I think I sleep, yust a little.

It seems likely that the close-up was taken out of continuity, allowing Ford to move his camera in on the actor. This creates an exciting cut which entails a pronounced sense of depth. In shot 4 (setup 2), Bennett moves away from the camera, resting his head back on the pillow, while reaching out to clasp Arrowsmith's hand (Fig. 2.35 – Fig. 2.36). It is followed by the closer view in which he speaks his final lines (Fig. 2.37). This angle of view could not have been shot simultaneously with the first and second setups because the third camera would have been visible in their angle of view. Thus, for shot 5 alone, Ford takes advantage of the spatial and temporal ubiquity typical of single-camera shooting. Nonetheless he shot and staged the rest of the scene with two cameras running simultaneously, holding on to the early sound practice that was on the verge of becoming obsolete. What advantages did this system ultimately provide?

By abjuring shot-reverse-shot and concentrating on Bennett's lines in the relatively close framings of shots 2 and 5, Ford allowed the actor the same opportunities that Colman had enjoyed in the kitchen scene but on a smaller shot scale. He could extemporize, pause as necessary, repeat himself, make interjections, and vary the volume, tempo and phrasing of his lines. In addition to multiple-camera shooting, the practice of mixing the sound live gave Ford and Bennett the added resource of the simultaneous recording of sound effects. The shooting script had called for silence in the last two shots, with images of the rain falling through the hospital roof and wetting Sondelius's face and body at the moment of death. In filming, Ford kept to the basic idea of no dialogue, but utilized sound rather than the image track to emphasize the rain. The film marks Sondelius's passing in shot 7 by the cessation of his energetic voice, and the continuation of the weather noise at high volume augmented by a fainter sound of weeping. In general, the rain sounds are quite pronounced in this scene and in the short scene outside the hospital which precedes it. The sound engineer's log complains of several takes "ruined by rain noise," suggesting the scenes were technically difficult to shoot and mix. It seems likely that the director was pushing against the

Figure 2.32: *Arrowsmith* death scene, shot 1.

Figure 2.33: *Arrowsmith* death scene, shot 2.

Figure 2.34: *Arrowsmith* death scene, shot 7.

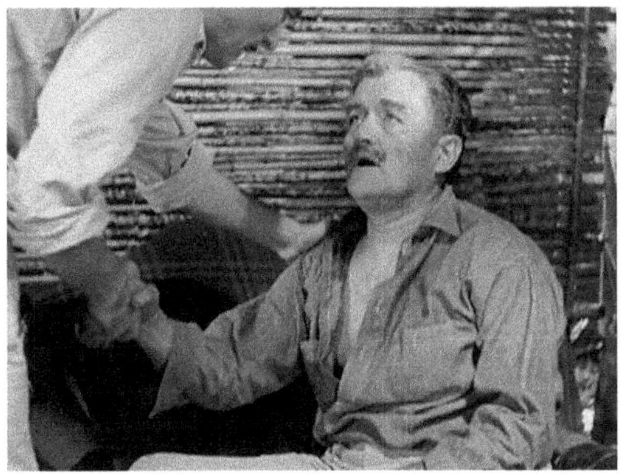

Figure 2.35: *Arrowsmith* death scene, shot 4a.

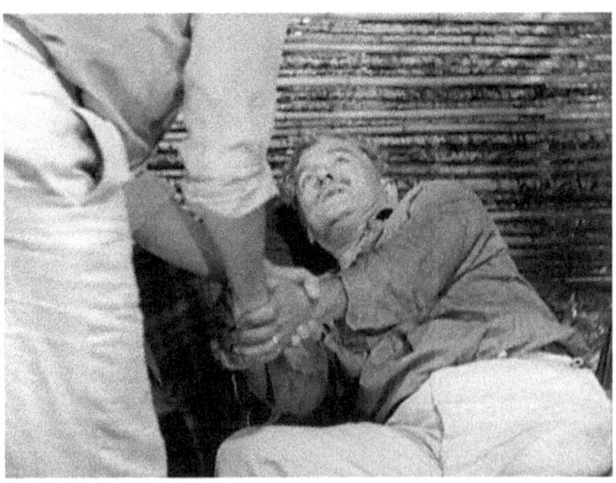

Figure 2.36: *Arrowsmith* death scene, shot 4b.

Figure 2.37: *Arrowsmith* death scene, shot 5.

limits of the technology in making the rain sounds as loud as possible, trying to keep them in the sonic foreground even as the actor spoke. Despite the difficulties, it was advantageous for the actor to be able to hear and react to the sound effects on set. Bennett could drop his voice when he was not very concerned to be heard, as he did in shot 4, and raise it when the words really mattered, as in shot 5. He could hear when the thunder started in shot 5 and pause to allow the crash to register before continuing his speech. The temporal integrity of the long take, and the simultaneous recording of voice, sound effect, and picture gave Bennett a chance to control and modulate his performance across the arc of the long takes.

We have tended to think about multiple-camera shooting as a clumsy but necessary stop-gap measure – a way of maintaining high cutting rates in a period when the camera was imprisoned in its blimp or booth. But the example of *Arrowsmith* suggests another way that filmmakers responded to the technical constraints and opportunities of the conversion. Ford, along with several of his colleagues, responded positively to the elimination of dialogue titles and the consequent demise of relatively fast-cut silent conversation scenes organized around intertitles. They took advantage of the conditions of early sound recording to go in the opposite direction: experimenting with static long takes. Indeed, the technical restrictions which favoured the production of a sound recording taken live without cuts may also have helped to inspire a similarly continuous visual record of the performance. The benefits of the long-take style in this context is perhaps most obvious in the case of the single-take scenes from *Seed* and *Arrowsmith* discussed here, but it also holds for the multiple-shot death scene of Sondelius which I have argued was structured to permit key moments of the performance to unfold, unhampered, in real time.

Bibliography

Arrowsmith Production Reports, Samuel Goldwyn Papers, Special Collections, Margaret Herrick Library, Academy of Motion Picture Arts and Sciences, Beverly Hills, California.

Bogdanovich, Peter. *John Ford*. Rev. ed. Berkeley: University of California Press, 1978. First published 1967.

Bordwell, David, Janet Staiger, and Kristin Thompson. *The Classical Hollywood Cinema: Film Style & Mode of Production to 1960*. New York: Columbia University Press, 1985.

Copeland, Peter. *Sound Recordings*. London: The British Library, 1991.

Gad, Urban. *Der Film: seine Mittel – seine Ziele* [1919]. Translated by Julia Koppel. Berlin: Schuster & Loeffler, 1921.

Jacobs, Lea. "The Innovation of Re-recording in the Hollywood Studios." *Film History* 24 (2012): 5–34.

Millard, André. *America on Record: A History of Recorded Sound.* Cambridge: Cambridge University Press, 2005.

Salt, Barry. "Barry Salt's Database." Accessed February 2, 2022. http://www.cinemetrics.lv/satltdb.php.

Salt, Barry. *Film Style & Technology: History and Analysis.* 3rd ed. London: Starword, 2009. First published 1983.

VanCour, Shawn. *Making Radio: Early Radio Production and the Rise of Modern Sound Culture.* New York: Oxford University Press, 2018.

About the Author

Lea Jacobs is the Mae D. Huettig and Vilas Distinguished Professor Emerita at the University of Wisconsin-Madison. She is the author of *The Wages of Sin: Censorship and the Fallen Woman Film* (1991), *Theatre to Cinema* (written with Ben Brewster, 1997), *The Decline of Sentiment: American Film in the 1920s* (2008) and *Film Rhythm After Sound: Technology, Music and Performance* (2015). Recent essays include "John Stahl: Melodrama, Modernism and the Problem of Naïve Taste," in *Modernism/Modernity* and "Making John Ford's *How Green Was My Valley*," in *Film History*. She is currently preparing the monograph *John Ford at Work: The Films in Their Production Context, 1928–1941.*

3. Expressive Visual Effects from Silent to Sound Film

Katharina Loew

Abstract: It is often assumed that with the introduction of synchronized sound, 'silent-style' visual effects largely disappeared as an expressive device. However, sound films continued to render abstract ideas by means of multiple exposure and split screen mattes. The resulting composites, which I call *montage shots*, do not seek to represent physical realities within the diegesis. Instead, they are clearly recognizable as formal devices that encourage viewers to forge conceptual connections between simultaneously presented images. The case of montage shots attests to a far greater consistency between silent and sound aesthetics than is usually acknowledged. It reveals how early sound films cultivated and expanded silent cinema's modes of expressivity, but also relegated them to discrete non-mimetic episodes. Attention to alternative approaches to expressive visual effects also allows us to rethink the instance of the Hollywood montage sequence, which became a staple of American filmmaking for decades to come.

Keywords: visual style, composite shots, montage sequences, special effects, expressivity

Early sound films are often considered visually uninteresting, largely dialogue-driven and shaped by requirements of sound recording equipment. 'Silent-style' visual effects – specifically expressive ones resulting from multiple exposure or combination printing – are assumed to have largely disappeared with the coming of sound. In actual fact, early sound films feature a remarkable range of expressive visual effects, which build directly on silent-era practices. In this chapter, I will focus on composites that visibly juxtapose different image components in one shot, either layered

Wiegand, D. (ed.), *Aesthetics of Early Sound Film: Media Change around 1930*.
Taylor & Francis Group, 2023
DOI 10.5117/9789463727372_CH03

on top of one another or presented side-by-side. Unlike sequential montage, which forges meaning from disparities between successive shots, what I call *montage shots* present diverse images simultaneously. Such composites are immediately recognizable as a purposeful arrangement and thus diminish the illusion of direct imprints of a physical reality. As I will show, montage shots persisted in sound cinema, where they facilitated enduring efforts to communicate ideas and emotions non-verbally. Highlighting the continuities between silent and early sound aesthetics, attention to montage shots allows us to revisit common assumptions about filmmaking in the early sound period.

Montage Shots in Silent Cinema

Already in the earliest days of cinema, filmmakers experimented with combining multiple scenes in one frame. Building on familiar iconography from religious art, lantern slides, illustrated journals, or advertising, early films juxtaposed different locations to represent projections of characters' 'inner eyes,' that is, their mental images or accounts. Further, portrayals of communication devices like film, television, magic lanterns, and mirrors motivated the combination of distinct spaces in the same shot. As Jan Olsson demonstrates, three-panel split screen composites served as the standard means for representing telephone conversations on European screens between 1906 and 1916.[1] Simultaneously, montage shots also became prevalent in contexts that were not motivated by character or technology. Instead, as the following examples suggest, they served as a self-conscious narrational gesture: Gaumont's *Le homard* (*Lobsters: All Styles*) (1913, dir. Léonce Perret) contains a three-panel split-screen shot. On the left, Suzanne, who is under the assumption that her husband Léonce is out at sea fishing during a stormy night, fearfully prays for his safety. The middle panel shows the rocks and sea, while on the right Léonce enjoys himself at the local movie theatre. The shot is tantamount to the ironic commentary of an omniscient narrator. It highlights the incongruous simultaneity of Suzanne's melodramatic response, the perilous seascape – suggestively shown devoid of any human presence – and Léonce's upbeat, mundane visit to the cinema. In Ernst Lubitsch's *Die Puppe* (*The Doll*) (1919), a montage shot humorously renders visible an auditory experience: twelve circularly masked moving mouths convey the clamour of Baron de Chanterelle's

1 Olsson, "Framing Silent Calls," 157–92.

legacy-hunting relatives. Finally, Fritz Lang's *Dr. Mabuse, der Spieler* (*Dr. Mabuse, the Gambler*) (1922), employs a montage shot to issue an ominous warning about Mabuse's power, ubiquity, and elusiveness. Following the protagonist's manipulation of the stock market, Mabuse's face, first in the guise of a trader and then as himself, is superimposed on the abandoned trading floor that is littered with the debris of the market turmoil he provoked. As Tom Gunning puts it, "the shot proclaims [Mabuse] the motivating force beyond this scene, the author appearing before the now empty stage."[2] As these examples suggest, montage shots were not only used to depict long-distance communication and character interiority, but they also rendered abstract thought and sensations associated with an omniscient narrational entity.

By the second half of the 1910s, cross-fertilization between compound aesthetics in cinema and modernist art became conspicuous. Films emulated the look of avant-garde photomontage, which in turn seized on many impulses from cinema. Dadaists, Surrealists, Cubists, and Futurists, strove to dynamize their work, devise motion effects, dismantle linear perspective, and create new levels of meaning through the compositing of image fragments. Modernist photomontages differ from their predecessors insofar as clearly comprehensible connections between image components became increasingly rare. Instead, they often encourage more visceral, associative modes of reception. In similar ways, cinematic montage shots evolved from presenting conventional spatial and narrative relations, for instance when simultaneously depicting characters and their 'thought bubbles,' to more complex and imaginative composites that seem more akin to mental associations and refrain from offering narrative meaning that can be easily verbalized.

Guido Seeber's Ideograms

One of the pioneers of highly sophisticated montage shots was the German cameraman and trick specialist Guido Seeber. His 1927 book *Der Trickfilm*, the first-ever book-length study on special effects, explains in detail how to split the image field with multiple irregular masks and how to devise a "technical score" to time the different image components. As Seeber lays out, montage shots constituted one of the principal "trick techniques of tomorrow," because they were capable of rendering abstract thought.

2 Gunning, *The Films of Fritz Lang*, 103.

Seeber's approach becomes apparent in his promotional film for Kipho, the 1925 cinema and photo exhibition in Berlin. The five-minute film, which Seeber co-created with advertising film producer Julius Pinschewer, was supposed to provide a "cross-section through the [medium] film at a furious pace," as Seeber puts it.[3] Approximately half of the film's roughly fifty shots were realized by means of split screen composites with three or more image components. Composites illustrate various aspects of film history and steps in the film and photo production process, referencing animation devices like the zoetrope and the praxinoscope, photo development, film drying and film editing. The concept of photography is portrayed in a shot that shows a photographer and her model with a potted plant on the top left and bottom right, a photographer underneath a dark cloth handling a view camera on a tripod on the bottom left, Königsberg castle turning on its head on the top right and a curved, white five-pointed star flashing repeatedly to signify the shutter in the centre of the frame. The principal visual motif of the film is rotational motion. Everything spins: the film reels, shutters, camera, and tripod cranks, intermittent mechanisms, rewinds, processing drums and praxinoscopes. The film's ubiquitous circular motion, which is intensified by the fact that the montage shots frequently juxtapose images turning at different speeds and in different directions, constitutes a verbal pun: The German verb *drehen* means both "to rotate" and "to film."

Inviting viewers to extract an idea from the juxtaposition of different image components, the Kipho film bears resemblance to picture puzzles. In fact, Seeber, in collaboration with director Paul Leni, and scriptwriter Hans Brennert, continued to experiment with film as visual language. In 1926/1927 they realized a series of eight roughly ten-minute-long crossword puzzle films that were advertised as "Rebus Film." Each puzzle was accompanied by a separately screened "solution" that provided additional clues before revealing the answers. Michael Cowan has rightly called attention to the fact that these films function quite differently from the eponymous puzzle devices that combine illustrations with individual letters to refer to homophonic words or phrases.[4] Unlike traditional rebuses, the films are not based on the sound of words and instead capitalize on the contemporary crossword vogue: they present visual cues for printed crossword puzzles that audiences solved while watching. Only an English-language version of the first episode survives. It offers a collage of stock footage, intertitles,

3 Seeber, *Der Trickfilm in seinen grundsätzlichen Möglichkeiten*, 246 (unless noted otherwise, translations are my own).

4 Cowan, "Moving Picture Puzzles," 197–218.

hand-drawn and object animation, as well as nine montage shots, each consisting of three to five image components. Montage shots do not play a major role as solving clues, which are primarily given in successive single shots of the same referent. Nonetheless, composites introduce the filmed crossword, provide viewer instructions, and also give some cues. Two triple composites in the solution film show various animated objects – dice, an abacus, matryoshka dolls, playing cards, a candle ring, and matches. Viewers are prompted to glean what the items have in common, namely that they refer to the number nine. The juxtaposition of concrete objects synthesizes an abstract idea.

Seeber's objectives must be seen in the context of modernist combinations of concrete images into new and often abstract forms. Cubists, for instance, strove to depict subjects from different points in space and time simultaneously, synthesizing multiple perspectives into a single image. Keen interest in compositing often coincided with the ambition to register and transmit thought through images. Taking inspiration from Egyptian and Chinese script, specifically their symbolic origins and compound nature, commentators touted ideography as a more immediate, universal, and multifaceted mode of expressivity. Ezra Pound's "ideogrammic method" involved juxtaposing concrete poetic imagery to express conceptual content. Some early observers also diagnosed an ontological kinship between cinema and ideographic writing. American poet Vachel Lindsay, author of the presumably earliest book-length film theoretical work, attempted to produce a fixed lexicon of cinematic hieroglyphs based on iconic, legible objects.[5] Precisely because silent cinema was perceived as tethered to the "concrete visual," (to borrow from Christian Metz) thinking in terms of ideography made the "abstract visual" seem more attainable.[6] The objective of rendering abstract thought through image synthesis also informed Sergei Eisenstein's notion of the ideogram, which he saw as the starting point for an "intellectual cinema."[7] In "The Cinematographic Principle and the Ideogram," Eisenstein writes:

> The point is that the copulation (perhaps we had better say, the combination) of two hieroglyphs of the simplest series to be regarded not as their sum, but as their product, i.e., as value of another dimension, another

5 Lindsay, *The Art of the Moving Picture*, 199–216; Gunning, "Vachel Lindsay," 19–30. See also Hansen, "Mass Culture as Hieroglyphic Writing," 42–73.

6 Metz, *Language and Cinema*, 275.

7 Pound, *ABC of Reading*, 26–28.

> degree; each, separately, corresponds to an *object*, to a fact, but their combination corresponds to a *concept*. From separate hieroglyphs has been fused – the ideogram. By the combination of two "depictables" is achieved the representation of something that is graphically undepictable.[8]

The ideographic process that Eisenstein describes here corresponds closely to Seeber's objectives for his montage shots. By juxtaposing different picture components in the same frame, he sought to articulate ideas not contained in each source image separately.[9] As we have seen, the ideograms that characterize the Kipho and the Rebus films are relatively simple and clearly legible. However, in subsequent feature films, particularly *Geheimnisse einer Seele* (*Secrets of a Soul*) (1926, dir. G. W. Pabst) and *Dirnentragödie* (*Tragedy of the Street*) (1927, dir. Bruno Rahn), Seeber employed substantially more complex montage shots to depict heterogenous sensory and mental processes. *Dirnentragödie* tells the story of Auguste (Asta Nielsen), an aging prostitute in love with a much younger man. Auguste plans a new life as a pastry shop owner with him, only to discover that during her absence he and her pretty young friend locked themselves into her bedroom. The following sequence condenses Auguste's anguish. Her downstairs neighbour plays the popular waltz "Wenn die Liebe nicht wär'" ("If Love Didn't Exist"), which Auguste had requested to accompany her anticipated engagement celebration. She breaks down in front of her locked bedroom door to the sounds of the neighbour's piano. The following forty-five-second-long sequence consists of thirty-six shots that suggest shreds of memories and auditory impressions that pummel Auguste during her mental collapse. The sequence contains both rapid editing and montage shots. The sequential montage alternates between Auguste on the floor, the neighbour on the piano, animated titles of her pleas for him to stop, her fists drumming on the floor and a series of nine, extremely brief shots of just a few frames each that summarize the events leading up to this moment. In addition, it features five montage shots that show piano keys, hammers and pedals, a female ear, the score of "Wenn die Liebe nicht wär,'" the word "Aufhören!" ("Stop!"), and eventually a close-up of Auguste. While the other image components revolve around acoustics, signalling a connection between ear, voice, and piano, Auguste's face and her direct look at the camera establishes an immediate link to the audience, as if granting direct access to her inner life. Further, transparent hands, failing to cover her face, allude to her desperation and shame (Fig. 3.1).

8 Eisenstein, "The Cinematographic Principle and the Ideogram," 29–30.

9 Seeber, *Der praktische Kameramann*, 240.

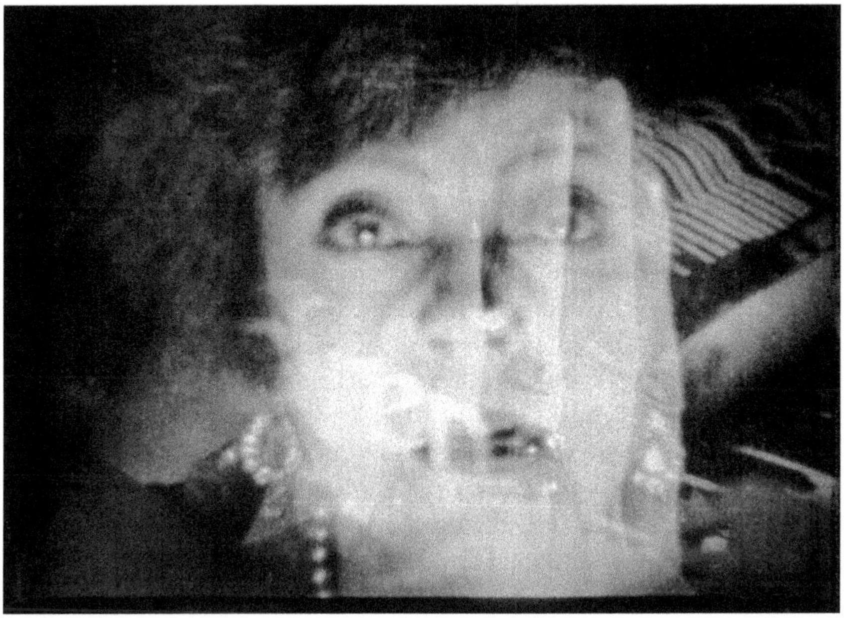

Figure 3.1: Montage shot in *Dirnentragödie* (*Tragedy of the Street*) (1927, dir. Bruno Rahn) (courtesy of Deutsche Kinemathek).

As the sequence demonstrates, in *Dirnentragödie*, montage shots are no longer limited to representing concepts. Instead, they can be described as polysemic displays that approximate a thinking in images that is apprehensible both through ideation and sensory perception. As Michael Cowan has noted, "Like Eisenstein's intellectual montage, Seeber's visual concepts occupied a middle ground between sensuous image and abstract thought."[10] Montage shots seek "to create *a new quality of the whole from a juxtaposition of the separate parts*," as Eisenstein put it.[11] Even though, as Yuri Tsivian has highlighted, "it is not so much shots per se that the montage director taps for meaning as cuts between shots,"[12] Eisenstein nonetheless employed montage shots at critical junctures in his films, as demonstrated by the closeup of the half-owl-half-human undercover agent in *Stachka* (*Strike*) (1925), a machine gun firing in all directions in *Oktyabr': Desyat' dney kotorye potryasli mir* (*October: Ten Days That Shook the World*) (1927), and the breeding bull towering in the clouds in *Staroye i novoye* (*Old and New*) (1929). In contrast to sequential montage, montage shots privilege

10 Cowan, "Moving Picture Puzzles," 204.

11 Eisenstein, "Dickens, Griffith and the Film Today," 238 (emphases in the original).

12 Tsivian, "Cyberspace and its Precursors," 90.

simultaneity with the goal, as Seeber writes in 1930, of achieving "total impression."[13] Because viewers are permitted to absorb the relations between montage units concurrently, montage shots tend to take on an iconic quality.

At the Angel Bar: *Der brave Sünder* (1931)

Contrary to what might be expected, montage shots persisted in sound film. To some extent, filmmakers continued to use montage shots in otherwise effect-free scenes to pointedly encapsulate an idea. Well-known examples for standalone montage shots in early sound film include the final shots of *All Quiet on the Western Front* (1930, dir. Lewis Milestone) and *Shanghai Express* (1932, dir. Josef von Sternberg). The former shows images of young soldiers, superimposed on a vast field of grave crosses, gazing back at the camera, while the latter superimposes a shot of the kissing couple on that of a busy train station, accentuating Madeline's ironic final line: "There is no one here but you and I." In *Les croix de bois* (*Wooden Crosses*) (1932, dir. Raymond Bernard), a victory parade is undercut by two shots of a superimposed procession of dark figures slogging along heavenward, and in *Das lockende Ziel* (*End of the Rainbow*) (1930, dir. Max Reichmann) Loisl's and Leni's first impression of the big city upon their arrival in Berlin is captured in a brief shot of superimposed neon signs.

Although expressive composites continued to be used as inserts, given that they represent a stark departure from the dominant tenets of verisimilitude that characterizes most dialogue films, they appear more commonly in longer, effect-heavy and/or sequential montage sequences. A particularly striking example can be found in Fritz Kortner's directorial debut *Der brave Sünder* (*The Upright Sinner*) (1931). Here, during a visit to a nightclub, the protagonist gradually descends into a hallucinatory state. The episode not only highlights the pertinence of expressive visual effects in early sound film, but also demonstrates how sound added a semantic level to cinematic compositing.

Der brave Sünder is, as Kortner explains, "set in the border area where tragedy and comedy meet."[14] Eminent character comedian Max Pallenberg gives a bravura performance as Leopold Pichler, a middle-aged head cashier from the Austrian provinces. The film indulges in Pichler's quirkiness and sharp-witted linguistic dexterity, while poking fun at his awkwardness and obstinacy. His beadledom is reflected in a stooped posture, scruffy little

13 Seeber, *Kamera-Kurzweil*, 226.
14 "Zwei Prominente debütieren," 3.

moustache, slicked-back, centre-parted and thinning hair, and pince-nez spectacles. At home, embittered by his emasculated existence as ineffectual head of an unruly family, Pichler indulges in barbs and violent fantasies. In the professional sphere, which he uncompromisingly isolates from his private life, Pichler is an officious stickler who derives his identity entirely from work. "Office is not a place. Office is an internal state," he advises his young assistant Wittek (Heinz Rühmann). Excessive attachment to standards and rules seems to be geared towards offsetting the chaos around him. Ironically, however, it is Pichler's own clownish ineptitude and disorientation that are particularly conducive to pandemonium. Oversized clothes, an out-of-place top hat, and emphatic gesticulation counteract his poignant struggle for dignity. Frantically clinging to a briefcase that embodies his dependability ("Auf Pichler ist ein Verlass"/"You can count on Pichler"), he becomes an object of pity and ridicule.

Early in *Der brave Sünder*, Pichler and Wittek travel to Vienna to deliver a large sum of money. Waiting for their boss, Director von Härtl (Ekkehard Arendt), at the Angel Bar, they are seated in the director's box and served champagne, which they consume first reluctantly and then more eagerly. The eighteen-minute-long Angel Bar sequence marks a crucial turning point: Pichler loses control of himself and the money in his custody, while his altered sense perceptions are conveyed through an abundance of camera, lighting, editing, and montage effects as well as (partly nondiegetic) musical accompaniment. The Black singer-dancer Kiddy (Rose Poindexter), who epitomizes unrestrained, primal corporeality – the very antithesis to Pichler – makes overtures to him. Her song "Your Volcan Eyes Really Do Tempt Me" becomes a musical salute to Pichler's imaginary virile and irresistible ego of bygone days. As he tells Wittek, he used to be a womanizing and gambling "double eagle with wings," a volcano, "a Stromboli." "Today," he adds wistfully, "I am a hill."

Over the course of the Angel Bar episode, Pichler's rigidity and repression make way for increasingly euphoric and extroverted behaviour, that includes affect displays like kisses and tears. On the dancefloor with Kiddy, he makes a mockery of himself and, to everyone's continued amusement, recovers the vitally important briefcase amid a Circassian dance troupe's dagger-throwing act. Kiddy, a blonde gold digger, the Circassian dancers, and a Russian baritone join the director's box; champagne and vodka flow freely, and everyone gets more and more intoxicated. Close-ups of Pichler begin to alternate with wider shots of the gathering in the box and medium shots/close-ups of two of the plaster putti that decorate the Angel Bar. As Pichler flirtatiously interacts with the putti (who now have human faces)

Figure 3.2: Montage shot in *Der brave Sünder* (*The Upright Sinner*) (1931, dir. Fritz Kortner).

to the tune of Kiddy's song, the shots of the box become scarcer, while the couples on the dancefloor move in slow motion: the outside world appears to recede.

To the solemn brass opening of the operetta aria "Ganz ohne Weiber geht die Chose nicht" ("No, We Can't Live Without Women, No"), a sentiment that Pichler earlier obstinately challenged, the putti take wing and float, superimposed, across the distorted room and his face (Fig. 3.2). Images and sounds of the present moment – like the waiters demanding the settlement of the exorbitant bill – coalesce with visual and auditory snippets from Pichler's private and professional lives. Ever more elaborate visual effects depict the merrymakers at the Angel Bar shrinking into the background as the general ledger glides across the screen and the office porter Kalapka appears in a Circassian outfit, proclaiming, "One can no longer depend on anybody, not even on Pichler." Kalapka throws a large knife past Pichler, who is standing at the service window of his office holding the bill, into the arms of Mrs. Pichler, who uses it to prepare a sandwich, a familial bone of contention earlier in the film.

To the popular Viennese folk tune "Oh, Du lieber Augustin, alles ist hin!" ("Oh, You Dear Augustin, All Is Lost!") a money roll is pulled out of the briefcase and coins are counted onto a surface labelled "coffer." Images of a fantastical locomotive with an iron face for a smokebox door, rolling eyes for headlights, and teeth for a cowcatcher are superimposed on a closeup of Pichler, and followed by a series of shots of pistons, wheels, tracks, and trains. Referencing their attempts to catch up with Director Härtl amidst a rainstorm and their trip to Vienna, Wittek and Pichler are now seen perched in the pouring rain on top of a moving locomotive holding a shredded umbrella. Pichler gets blown off, flies backwards through the air, and ends up clutching on to the finial of a church tower. This phallic imagery is amplified in the soundtrack by the sexually suggestive folk song "The Hunter from Kurpfalz" and Papageno's "A Girl or a Woman" from Mozart's *The Magic Flute*. As the tower tellingly topples, Pichler falls on the train tracks and is run over. After the train has passed, Pichler sits up, mutters "too bad about me," and sinks back on the tracks. As the dream fades away, he finds himself fully dressed and still clutching the briefcase, in Kiddy's bed.

The sequence is particularly noteworthy for the way it renders the gradual process in which the protagonist dissociates from outside reality. It offers a believable representation of how people experience the process of inebriation. Images and sounds – past, present, and imaginary – flow into each other and culminate in a fanciful dream that never entirely severs its ties to the real. Rudolf Arnheim described the scene in enthusiastic words:

> The way there is dream where there is still reality; the way there is reality where there is already dream. The way an earthly room dissolves in spotlights into a ghostly flare, the way a thoroughly believable dance hall turns into Hell, where the petit-bourgeois symbols of vice and crime gain substance – 'a sensuous trick, a trick of the senses,' as Pallenberg stammers. The naked Negress dances, knives whiz into the holy briefcase, and very slowly, in clever mirror and slow-motion shots, the world blurs; very slowly it becomes a dream, the most beautiful dream that film art has produced. Neither in *Secrets of a Soul* nor in *Narcosis* was anything close to this accomplished. Here neither fantasy nor the applied arts rule; the precise illogic of the real dream picks each motif from reality, mixes up that which does not belong together, softens the real relationships.[15]

While not all commentators responded as ardently as Arnheim, many agreed that the sequence constituted a showpiece of film art. Journalist Hans Wollenberg lauded that "Pallenberg's intoxicated daydream demonstrated in

15 Arnheim, "Partly Expensive, Partly Good," 177. The films referred to are *Geheimnisse einer Seele* (1926, dir. G. W. Pabst) and *Narkose* (1929, dir. Alfred Abel).

unprecedented perfection [...] that film – and only film – is granted access to such psychological depths."[16] Author Otto Alfred Palitzsch wrote: "Kortner occasionally transcends the border between filmic and literary ideas. The camera no longer sees, it thinks. [...] Through this, his first work, Kortner has established himself as one of the best sound film directors worldwide."[17]

Although significantly longer and more elaborate, the Angel Bar sequence shares important technical and thematic concerns with Auguste's breakdown in *Dirnentragödie*. Both blend montage shots, transitional effects, and sequential montage to express an individual's sense perceptions, memories, and altered states of mind. Nonetheless, *Der brave Sünder* compounds Seeber's compositing technique by employing fragments of speech (such as the waiters' repeated "the bill, gentlemen, the bill!") and music, often expounded by implicit or explicit lyrics, to create associative links and tensions with the imagery. As the Angel Bar episode demonstrates, *Der brave Sünder* approaches sound in ways that are consistent with silent era stipulations for film art, namely the effort to prevail over unoriginal reproductions and instead employ cinema's technological tools for creative ends. By fusing picture montages with phonomontages, it set standards for film art in the era of synchronized sound.

Hollywood, Montage, and the Hollywood Montage Sequence

For lack of another term, observers may be inclined to describe Auguste's breakdown in *Dirnentragödie* and the Angel Bar sequence in *Der brave Sünder* as "montage sequences." Though technically plausible, the label also seems unsuitable, as it commonly designates a specific and widely used tool that has become a cliché in film narration. David Bordwell and Kristin Thompson define the montage sequence as a "device for temporal ellipsis" in which "brief portions of a process, informative titles (for example, "1865" or "San Francisco"), stereotyped images (such as the Eiffel Tower), newsreel footage, newspaper headlines, and the like can be joined by dissolves and music to create a quick, regular rhythm and to compress a lengthy series of actions into a few moments."[18] Neither *Dirnentragödie* nor *Der brave Sünder* aspires to omitting a portion of the sequence of events. Instead, each constitutes a critical juncture that seeks to capture a character's mental

16 Wollenberg, "Der brave Sünder."
17 Palitzsch, "Der brave Sünder."
18 Bordwell, Thompson, *Film Art*, 252.

life in all its facets, including sentiments, thought fragments, and sense perceptions. They prompt us to re-examine common assumptions about expressive visual effects and the stereotypical Hollywood montage sequence.

According to Bordwell, the Hollywood montage sequence established itself at the outset of the sound era and transposed expository information, which a silent film would have imparted through intertitles, into symbolic imagery:

> Functionally, the montage sequence can be thought of as the rhetorical amplification of an implicit intertitle: 'The world erupts in war,' or 'After two years in prison – ' or 'Her singing career rapidly declined.' In this respect, the silent film had already mapped out a role for montage sequences. By the end of the silent era, many American feature films were inserting footage depicting the sort of stereotyped symbols that might also illustrate an 'art title': a turning clock dial, wafting calendar pages, and so forth. The sound cinema began to eliminate expository intertitles and quickly expanded symbolic passages into full-fledged sequences. Shots, often of an emblematic or universalizing nature, would be linked by dissolves or superimpositions in order to cue a temporal ellipsis, phases of a repeated process or a general state of affairs. The montage sequence could be transitional or expository establishing a period or locale.[19]

While Bordwell may be justified in locating the emergence of the stereotypical Hollywood montage sequence at the dawn of the sound era, it should be noted that emblematic summaries of narrative events were already a staple in silent cinema. For instance, at the end of the sixth instalment of the Albatros serial *La maison du mystère* (*The House of Mystery*) (1923, dir. Alexandre Volkoff), an intertitle describes the protagonists' contributions to the war effort. Subsequently, World War I is summarized in sixty seconds: ominous storm clouds are followed by a series of eight shots of combat operations and a victory parade at the Arc de Triomphe in Paris. All images are linked by dissolves and the years 1914 to 1919 are successively superimposed on the centre of the frame. In function and style, this sequence closely resembles the stereotypical Hollywood montage sequence of the sound era. However, scholars have largely overlooked such instances and instead pointed to expressive sequences as precursors. Barry Salt, for instance, claims that *Die Straße* (*The Street*) (1923, dir. Karl Grune) "contains one of the very first fully realized montage sequences in the classical form, in which the dissolves take

19 Bordwell, *Narration in the Fiction Film*, 186–87.

place absolutely continuously, so that there is always a changing sequence of superimposed images present on the screen."[20] Yet, the episode is far from symbolically synthesizing a narrative development. Instead, as Anton Kaes lays out, it provides essential insights into the psyche of the protagonist, perceptions that are difficult to verbalize:

> In a film-within-the-film insert, demarcated as a daydream, the film shows us what he [the protagonist] imagines seeing: cars and trains in wild motion, thrill-seekers enjoying themselves on fairground rides, a circus clown making faces, an organ-grinder. The final image, held longer than all other shots, is of a smiling young woman who beckons him with her yes. Compressed within a frame, these overlapping vignettes of the city produce, like a cubist painting, multiple perspectives, temporal simultaneity, and dizzying dynamics.[21]

While there is little doubt that sequences that seek to condense space, time, and other narrative information became a staple in Hollywood filmmaking in the early 1930s, the device did not necessarily imply the use of elaborate special effects. In Hollywood, the term "montage" caught on around 1930 in response to writings by Eisenstein and Pudovkin. Its precise meaning, however, remained somewhat of a mystery, prompting cinematographer Karl Freund to remark in 1934, "the word 'montage' is probably the most maligned – and certainly the least understood – of any used by English-speaking film-people."[22] Following practices in Soviet film, montage was for instance often interpreted as indicating very fast editing.[23]

The use of expressive composites had been popularized in Hollywood in the second half of the 1920s by German films such as *Metropolis* and European filmmakers like F. W. Murnau, Paul Leni and Paul Fejos. This type of effect also became associated with the term "montage." As Herb Lightman writes in *American Cinematographer* in 1949, "there are three main styles [of montage sequences]: the cut montage, the dissolve montage, and the superimposed montage."[24] Throughout the 1930s and 1940s, the nature and significance of "montage" was regularly thematized in the American trade press. As this shows, "montage" was by no means valued as a device

20 Salt, *Film Style & Technology*, 194.
21 Kaes, "29 November 1923, Karl Grune's *Die Straße*," 124.
22 Freund, "Just What is Montage," 204, 210.
23 Ibid.
24 Lightman, "The Magic of Montage," 361.

for time compression, but instead it was championed as a tool to represent interiority, as Lightman emphasizes:

> But the potentiality of the montage goes far beyond the mere utilitarian expedient of condensing time or action. In itself, properly used, it is a potent dramatic device peculiarity suited to the scope of the cinematic medium. It is capable of conveying to the audience certain subtleties of mood and characterization which could not be as aptly portrayed in any other way. Used subjectively, for example, it can present a situation as it appears to a specific character in the story. [...] In a sense, the camera "crawls inside the mind" of the character. [...] In America, the art of creative montage has not yet been fully developed. The tendency is to fall into formulas of *cliché*, which achieve transitional purposes with little originality. Correctly used, montage approaches pure cinema – which after all is nothing more than the art of visually advancing a dramatic story without calling attention to the mechanical effects involved.[25]

Emphatically endorsing silent modes of expressivity, Lightman advocates the montage sequence as a medium-specific means for rendering interiority. However, expressive devices like montage shots constituted a major stylistic challenge for dialogue film, which largely embraced the pretence of rendering direct and unaltered imprints of a physical reality. As Rudolf Arnheim explains, "in a naturalistic narrative film a superimposition by double exposure may easily give the impression of a foreign body, which interferes with the style of the rest."[26] Thus, compared to silent films, the distinction between mimetic and mental/expressive realms tends to be clearer in sound films, at least in talkies. Hollywood films rendered this differentiation quite sharply, highlighting the marked otherness of the montage sequence. Precisely because abstract or mental subject matter was perceived as a "foreign body," it had to be sequestered from the mimetic main body of the film. In European cinema, in contrast, interiority and self-conscious narration tended to integrate more organically.

As I have suggested here, our understanding of the function and physiognomy of expressive visual effects in early sound cinema and the so-called "montage sequence" is still quite limited. Attention to montage shots complicates prevalent assumptions about the nature of non-mimetic, effect-heavy sequences and reveals crucial, yet often overlooked continuities

25 Ibid.
26 Arnheim, *Film as Art*, 121.

between silent and sound cinema. Following important work by Bordwell and Thompson, scholars have discussed composite devices almost exclusively in the context of Hollywood as concise illustrations of pertinent plot information. However, the emphasis on stereotypical uses in Hollywood films has hindered exploration of alternative approaches to expressive visual effects in and outside of the American film industry. While expressive visual effects disrupt the verisimilitude of dialogue film, they have remained important tools for depicting interior states. They allowed filmmakers to preserve, in dialogue film, something of what Herb Lightman called "the magic of pure cinema."

Bibliography

Arnheim, Rudolf. *Film as Art.* Berkeley: University of California Press, 1957.

Arnheim, Rudolf. "Partly Expensive, Partly Good" [1931]. In *Film Essays and Criticism.* Translated by Brenda Benthien, 173–79. Madison: University of Wisconsin Press, 1997.

Bordwell, David. *Narration in the Fiction Film.* Madison: University of Wisconsin Press, 1985.

Bordwell, David and Kristin Thompson. *Film Art: An Introduction.* New York: McGraw-Hill, 2019.

Cowan, Michael. "Moving Picture Puzzles: Training Urban Perception in the Weimar 'Rebus Films.'" *Screen* 51, no. 3 (Autumn 2010): 197–218.

Eisenstein, Sergei. "The Cinematographic Principle and the Ideogram." In *Film Form: Essays in Film Theory,* edited and translated by Jay Leyda, 28–44. New York: Harcourt, Brace & World, 1949.

Eisenstein, Sergei. "Dickens, Griffith and the Film Today." In *Film Form: Essays in Film Theory,* edited and translated by Jay Leyda, 195–255. New York: Harvest Book, 1949.

Freund, Karl. "Just What is Montage." *American Cinematographer* 15, no. 9 (1934): 204; 210.

Gunning, Tom. "Vachel Lindsay: Theory of Movie Hieroglyphics." In *Thinking in the Dark: Cinema, Theory, Practice,* edited by Murray Pomerance and R. Barton Palmer, 19–30. New Brunswick, NJ: Rutgers University Press, 2016.

Gunning, Tom. *The Films of Fritz Lang: Allegories of Vision and Modernity.* London: BFI, 2000.

Hansen, Miriam. "Mass Culture as Hieroglyphic Writing: Adorno, Derrida, Kracauer." *New German Critique,* no. 56 (Spring/Summer 1992): 42–73.

Kaes, Anton. "29 November 1923: Karl Grune's *Die Straße* Inaugurates 'Street Film,' Foreshadows Film Noir." In *A New History of German Cinema*, edited by Jennifer M. Kapczynski and Michael D. Richardson, 124–28. Rochester, NY: Camden House, 2012.
Lightman, Herb A. "The Magic of Montage." *American Cinematographer* 30, no. 10 (1949): 381–82.
Lindsay, Vachel. *The Art of the Moving Picture* [1915]. New York: Liveright Publishing, 1970.
Metz, Christian. *Language and Cinema*. Translated by Donna Jean Umiker-Sebeok. The Hague: Mouton, 1974.
Olsson, Jan. "Framing Silent Calls: Coming to Cinematographic Terms with Telephony." In *Allegories of Communication: Intermedial Concerns from Cinema to the Digital*, edited by John Fullerton and Jan Olsson, 157–92. Bloomington: Indiana University Press, 2005.
Palitzsch, Otto Alfred. "Der brave Sünder." *Vossische Zeitung*, October 23, 1931.
Pound, Ezra. *ABC of Reading*. New York: New Directions, 1934.
Salt, Barry. *Film Style & Technology: History and Analysis*. London: Starwood, 1983.
Seeber, Guido. *Kamera-Kurzweil. Allerlei interessante Möglichkeiten beim Knipsen und Kurbeln*. Stuttgart: Union Deutsche Verlagsgesellschaft, 1930.
Seeber, Guido. *Der praktische Kameramann. Theorie und Praxis der kinematographischen Aufnahmetechnik*. Berlin: Verlag der Lichtbildbühne, 1927.
Seeber, Guido. *Der Trickfilm in seinen grundsätzlichen Möglichkeiten* [1927]. Frankfurt am Main: Deutsches Filmmuseum, 1979.
Tsivian, Yuri. "Cyberspace and its Precursors: Lintsbach, Warburg, Eisenstein." In *Transmedia Frictions: The Digital, the Arts, and the Humanities*, edited by Marsha Kinder and Tara McPherson, 80–99. Berkeley: University of California Press, 2014.
Wollenberg, Hans. "Der brave Sünder." *Die Lichtbild-Bühne*, October 23, 1931.
"Zwei Prominente debütieren." *Mein Film*, no. 304 (1931): 3.

About the Author

Katharina Loew is Associate Professor of German and Cinema Studies at the University of Massachusetts Boston. She is the author of *Special Effects and German Silent Film* (Amsterdam University Press, 2021). Her work on silent cinema and film technology has also been published in *New German Critique, Film Criticism, Early Popular Visual Culture* and several edited collections. She is series editor for Cinema and Technology at Amsterdam University Press.

4. From the Lexigraphic to the Melomanic

Accommodations to Sound in American Studio Animation

Donald Crafton

Abstract: *Lexigraphic* and *Melomanic* are concepts proposed to fine tune our understanding of American studio animation's transition to sound. In the 1910s and 1920s, animators and audiences were accustomed to following cartoon stories as a kind of reading, assisted by icons and words-in-the-image stylistics adapted from comic strips and comic books. These symbols and picture-words anticipated *emoji*, delivering an idea, concept, or emotion in pictorial form. Many animators employed these image-texts and picture icons imaginatively and reflexively. Led by the competitive example of the Walt Disney studio when it converted to sound, the established lexigraphic mode of production and reception gave way quickly to melomania – an obsessive preoccupation with syncing screen action to a pre-recorded soundtrack.

Keywords: animation, writing in film, film music, icons, intermedial contexts

Lexigraphic and *Melomanic* – I invoke these two words to describe concepts that seldom have been used in film- or early-sound-film studies, but which may fine tune our understanding of animation history. The idea is really very simple. As I reviewed dozens of studio-made cartoons released between 1917 and 1932, I observed that their path to sound took an idiosyncratic turn. Cartoons in the silent period were a lexigraphic[1] medium. They habitually

1 From the Greek λέξις (lexis, diction, word). I use it to mean a visual display of words, discursive signs, or similar.

Wiegand, D. (ed.), *Aesthetics of Early Sound Film: Media Change around 1930*.
Taylor & Francis Group, 2023
DOI 10.5117/9789463727372_CH04

plastered writing, graphic expressions, and picture-word precursors of emoji onto their screens as genre-defining practices. Audiences absorbed these texts as a kind of reading. After the transition to the talkies, led by innovations from the Walt Disney studio, animators gradually abandoned this mode of production and instead foregrounded novel synchronization of music- and image-tracks. The obsession with music was a melomania.[2] Michel Chion touched on the phenomenon:

> The silent cartoon began its existence well before the popularization of radio and sound film. It even had the luxury of representing sounds visually, using the same codes as comic books. [...]
>
> The result [of the introduction of audio-visual synchronism] was the sort of film [...] where we see a cartoon creature sing and make music with anything, hitting this or blowing on that. The whole world becomes a wind, string, or percussion instrument creating a jazzy music that sets everyone and everything to dancing.[3]

It is obvious that cinema and comics have drunk from the same pool of discursive devices from the "lightning sketchers" at the beginning of film until now (the Marvel franchise, etc.). Animators and their studios, however, routinely modify existing conventions, bending them to consumers' expectations and the medium's capabilities. That was the distinct break in the conceptual framework of animation, away from a lexigraphic mode and toward a mode exploiting the compulsive-obsessive application of, and passionate appreciation of music, the melomanic.

Hieroglyphs without Sounds

Bobby Bumps Starts for School (1917, dir. Earl Hurd)[4] illustrates pre-sound cartoons' reliance on written words, texts, and discursive signs to tell their stories and generate humour. Many of the film's "codes" (Chion), however, were not derived from comic strips, but were shared with the live-action cinema of the day, for example, the practice of narrational and dialogue intertitles, of which *Bobby Bumps* contains several. There are also instances

2 From the French *mélomanie*: a musical fascination that becomes an obsession.

3 Chion, *Film, a Sound Art*, 40.

4 Available as *Bobby Bumps Starts to School*, DVD, Stathes Collection/Thunderbean Animation, *Cartoons on Film*, 2014.

of what Tom Conley calls *film hieroglyphs*, "the presence of alphabetical and iconic writing within the field of the moving image. Writing is never the film itself, but both the image and script are together, within each other, in various degrees of tension that cannot be resolved."[5] His description of scenes that, in mainstream filmmaking, contain, either accidentally or intentionally, signs and writing within the mise-en-scène that amplify its meaning or comment ironically on it, also applies to silent cartooning.[6] In pre-sound animated cartoons there are similar unresolved tensions between written texts and the images in which they appear. The hieroglyphs in cartoons are always intentional, since everything in the image is constructed. In *Bobby Bumps*, for example, there is an insert shot of Bobby's handwritten history assignment to show his ineptitude as a pupil (bad penmanship, spelling, and grammar) and to make a joke.

Many of these textual images are also *toponyms*, the graphic equivalent of a figure of speech used to place the time and place of a scene, explain the context, or provide in a picture a statement of what we need to know about the story, thus establishing the 'world' of the film.[7] *Bobby Bumps* begins with a title card "The Call of the Wild," which dissolves into a toponym – a bell ringing with the word "School" emblazoned on it. Then, "The Answer" (title card) shows Bobby's mother grooming him for his journey to the "wild" school. These texts provide story background, set an ironic tone, and eliminate the need for time- and labour-intensive animated expository scenes. They also specify abstractions, like the "wildness" of a school day, that might have been difficult to express as animated visuals.

Animation's use of these figures and conventions shared by live-action cinema suggests that the producers (the Bray Studio, in Hurd's case) were pursuing a double strategy of using live-action devices to assimilate their short films into "an evening's entertainment" format, while also trying to cultivate a distinctive 'look and feel' to distinguish their product. Deploying the familiar vernacular of comic strips was part of this strategy.

Bobby Bumps has its fair share of comics-related devices that were becoming endemic to classic animation cinema. Most of these were not, however, what we would call acoustic codes. Take *sight lines*, for example, which are dotted lines that connect a character's gaze to its object. These are

5 Conley, *Film Hieroglyphs*, xxiv.

6 I use "cartoon" and related words to refer to studio animated shorts and their creators. "Comics" and related words will refer to print-media comic art, strips, and books, and their creators.

7 Conley, *Film Hieroglyphs*, xxviii.

redundant because we already see that character is looking at something. These sight-line devices don't imply any sounds.

Punctuation marks were another cartoon feature with comic-strip analogues. Animators adapted these hieroglyphs, or "icons" (Scott McCloud's term),[8] as dynamic components of the cartoon genre. Felix the Cat led the way, as the crafty feline routinely turned icons representing his emotional state, including the exclamation points and question marks that appear overhead, into utilitarian objects: canes, eyeglasses, ladders, bicycles, and the like.

Linguists are intrigued by these unspoken components of written language.[9] Punctuation marks are not voiced; yet, they structure writing to conform to patterns of speech. Cartoons exploited the ambiguity produced in silent films when language that could not be spoken audibly had to have a visual equivalent. Thus, cartoons used the marks as quasi-linguistic expressions of moods, emotions, and the senses: "*?*" stands in for "huh?" (puzzlement) and "*!*" for "wow!" (astonishment). A scene in *Bobby Bumps*, for example, shows the teacher growing question marks out of his head to convey his perplexity as he tries to understand why his school bell has gone silent, but they don't imply specific words or sounds. There was an added advantage: viewers did not have to be literate or competent in English to understand the meanings of the marks.

Sound-Suggestive Hieroglyphs

Several comics codes were intended to represent sounds, and perhaps to invoke them. Take those lines known as *zip ribbons* that follow the trajectory of a moving object, often a person's fist.[10] In comics, these graphic conventions are not simply diagrams, they also represent the sounds of air rushing in to fill a vacuum, swishes and, of course, zip sounds. Sometimes they are paired with specific words that prompt us to 'hear' the sounds ("Whoosh!"). Because cartoon scenes show the actual duration of a motion, though, zip lines, such as the ones throughout *Bobby Bumps*, are redundant. There was no need for animators to anchor an object's rapid movement to a graphic convention like a zip, because audiences could just see the flying fist, or

8 McCloud, *Understanding Comics*, 27.

9 For a tongue-in-cheek demonstration of spoken punctuation, play Borge, "Phonetic Punctuation." For a serious linguistic appraisal, Fischbach, "Punctuation (1): The Linguistic Side."

10 McCloud, *Understanding Comics*, 110–11.

whatever. They usually were embellishments, a surplus effect. Nevertheless, cartoonists did it anyway! Indeed, there were few silent cartoons that *did not* add straight or curved parallel lines to signal an object's movement. Thereby, audiences had the option of imagining the whooshing sounds that would have accompanied these actions. We don't know how often viewers imagined or 'heard' the implied sounds, or even how many viewers took notice of the zip ribbons' existence. Motion lines' extraneousness notwithstanding, they remained a generic component of animation throughout the 1920s and into the sound period.

Impact effects that graphically adorn object collisions have the same ontology as motion lines. Again, in comics, a fist smashing a face might be shown as a zip ribbon leading to the face, accompanied by shock waves, stars, and other visible aftereffects. There might also appear in the panel an impact word, like "*BAM!*" or "*KRAK!*" In cinema, as was the case with motion lines, these graphic additions to impacts were unnecessary: The fist hits the face in real time and its effect is immediately visible, so no textual supplement is necessary. Nevertheless, animators dipped into comics vocabulary and used non-verbal icons to amplify the violent imagery. Such impact signs often were set in motion, producing rotating stars, surging dust clouds, or flying droplets of sweat and spit. Were these hieroglyphs also acoustic effects? Perhaps. As 1920s audiences responded with surprise, shock, or slapstick laughter, they may have imaginatively provided the sounds of the splashes, crashes, and booms. We don't know.

Several of the effects described above are *onomatopoeia*, words that, when uttered, produce their originating natural sounds. In *Bobby Bumps*, the words "Ding" and "Dong" included in the image give us the sound of the ringing school bell. We see the word "Hoo" to indicate the cat is gagging. *Scents and Nonsense* (1926, dir. Bill Nolan),[11] plays with the convention. Krazy Kat sprouts exclamation points and question marks when he learns of a dance contest. A sleeping man's "ZZZZZ" mimics snoring. These letters morph into an image of 'sawing logs,' argot for snoring and sleeping, and then into a surreal non sequitur: frankfurter sausage links, i.e., 'hot dogs,' that start barking, their action reinforced by "Woof!" Elsewhere, a bear voices an onomatopoeic "GRRR." Krazy grabs the letters and transforms them into a bicycle. A sound hieroglyph becomes reified as an object, and the daft cat neutralizes the acoustic sign of the charging bear's violent threat.

11 *Krazy Kat in Scents and Nonsense*, DVD, Stathes Collection/Thunderbean Animation, *Cartoons on Film*, 2014.

Characters in live-action silents often spoke without benefit of dialogue intertitles. Whether billet-doux, rants, or vulgarities (which audiences were eager to lip-read), these 'mute' articulations were an important component of pre-sound movies. Similarly, in silent animation there was a lot of *jabbering* mouth movement without benefit of either intertitle or balloon. These moments ask us to rely on the context to understand what the character is saying.

Word balloons and *thought balloons* were the most acoustically specific hieroglyphs in silent cartoons and conveyed talking and thinking. Will Eisner explains that, in comics, "The balloon is a desperation device. It attempts to capture and make visible an ethereal element: sound."[12] Animators, too, used balloons to convey sounds, but, one does not sense that they were desperate; rather, it's more likely that using word balloons was habitual, a device whose interpretation and function were familiar to artists and audiences, and pragmatic, a solution to the problem of needing to convey spoken information to illuminate a character's frame of mind, to condense a narrative, or to make a joke. (Some animators, including Hurd, were also comics artists; many were not.)

Word balloons question the ontological status of the 'voice' in silent animation. Chion has memorably developed the idea of *acousmatic* sound, "a word of Greek origin, applied here to signify *sound that one hears without seeing its cause*."[13] But isn't the word balloon just the opposite – *anacousmatic*? We do not actually hear the speech in these balloons, but we see its cause, the speaker. Perhaps Chion would explain this as *rendering* (which parallels what McCloud calls *closure*). Sound-images in silent films, like guns firing, and live sound-effects, such as those provided by pit-orchestra accompanists who mimicked the clip-clops of horses' hooves, were not imitations of actual sounds. On the contrary, the effects "create in the spectators' minds the very idea of a horse running fast."[14] Similarly, hieroglyphs in silent animation may be not literal, but instead, are logocentric, depending on us viewers to impute a character's voice (imagining, for instance, what the voices of Bobby, his dog, and his cat might sound like when each shouts "Towel!" in word balloons).

Many of the cartoons' hieroglyphs, as well as the balloons and other comics icons that developed over the previous century, resemble *emoji*. According to Marcel Danesi, "the definition of *emoji* is, simply, a 'picture-word' – a rather

12 Eisner, *Comics & Sequential Art*, 26.

13 Chion, *Film, a Sound Art*, 40 (emphasis in the original). Brian Kane refers to: "the experience of the acousmatic voice, a speaking voice whose source remains underdetermined" (Kane, *Sound Unseen*, 187).

14 Ibid., 7.

accurate characterization of what an emoji is."[15] These ubiquitous smileys, pictograms, ideograms, and hundreds of other signs have exploded over the past dozen years and have roots in Japanese manga comics. They are like words, having agreed-upon meanings. Like punctuation marks, they are nonverbal nuggets of subjectivity, communicating feeling, attitudes, and emotions that are difficult to express in short electronic communications. Retrospectively, many of the icons of comics and cartoons functioned as emoji *avant la lettre*. Krazy Kat, when he contemplates ways to kill himself in *Scents and Nonsense*, expresses his thoughts within a balloon over his head picturing a pistol, then a noose. Many cartoon icons now have cognate emoji, for instance impact effects (collisions), snoring, money bags, dust clouds, sweat drops, stars, hearts, and punctuation marks. Emoji, Danesi continues, "are stylized, almost comic-book-like, pictures that can (and do) replace words and phrases. Their main function seems to be that of providing nuances in meaning in the tone of the message. So, they are not completely substitutive of traditional written forms; rather, they reinforce, expand, and annotate the meaning of a written communication."[16] For a specific comparison to animation, consider the emoji *Music Note* 🎵 and *Musical Notes* 🎶 adopted in Unicode 6.0 in 2010, and defined as "song lyrics, or other music related topics."[17] Analogously, music-note icons in cartoons accompany instruments, bands, singing and whistling voices – music-emitters of any kind. Nolan, in *Scents and Nonsense*, bizarrely has Krazy Kat remove his ears (organs of hearing) to use as drumsticks (sound producers), with the drum's beat then visualized as flying emoji-notes.

Are these cartoon emoji depicting diegetic anacousmatic music, that is, sound emanating (though inaudible) from on-screen sources, or do they signify the audience's experience of hearing? Or both? *Bright Lights* (1928, dir. Disney),[18] one of innumerable examples, features an orchestra producing swarms of animated notes showing the sound of playing, as well as the audience's hearing. The notes are not onomatopoeia because, although the depictions are specific notation values,[19] they really represent the 'idea' of music (following Chion), not a particular sound or tune. We might regard them as hieroglyphs – or emoji – expressing, *now there is music*.

15 Danesi, *Semiotics of Emoji*, 2.

16 Ibid., 15.

17 https://emojipedia.org/musical-note/.

18 *The Adventures of Oswald the Lucky Rabbit*, DVD, Walt Disney Home Video, 2007.

19 The Musical Note emoji: "Displays as two eighth-notes (*quavers* in British English) connected with a beam. A single eight-note (*quaver*) is used in Microsoft and Google's emoji designs" (Emojipedia "Musical Note").

Intended or not, music accompanists in cinemas may have picked up on these and other cartoon emoji as cues for performing improvised music, as well as adding spikes, or other sound effects. Which, again, asks us to question where the 'sound' originates. It seems to issue from the screen actors and environment (diegetically), from the orchestra pit, from the intentions of the animator, and is generated in the imagination of the film viewers. The answer appears to be that it is a function of the collaborative performance that is animation cinema, dependent on interconnected systems of production and reception, and that its function is cultural as well as technological.

Hieroglyphs and the Coming of Sound

The story often has been told of how Walt Disney and Ub Iwerks designed their Mickey Mouse character, added sync sound, and achieved success with *Steamboat Willie* (1928, dir. Disney).[20] The film launched, in Iwerks' phrase, "the sound renaissance of 1928."[21] Here are the roots of animation's melomania. The celebratory accounts of synchronizing cartoon action with recorded music and sound effects have not dwelled on what was left behind, namely, silent animation as a lexigraphic medium liberally speckled with hieroglyphs.

Disney's staff innovated tools for pre-planning the synchronization of image and sound in bar sheets and exposure sheets, which were road maps for the audio-visual course of each film. This counter-intuitive strategy locked in the sound in advance of the animating. Disney sound-man William Garity wrote:

> This one phase of producing cartoons is probably the least understood by the public, although it is perhaps the simplest part of the problem. Since the advent of talking pictures and the standardization of film speed, the problem became simply one of resolving all musical tempos in terms of the standard speed, and of making a consecutive series of drawings to fit this tempo.[22]

The process in the 1930s at Disney and the other studios (even Fleischer's, which resisted at first) was to develop the complete film as a script and/or

20 *Walt Disney Treasures: Mickey Mouse in Black and White,* DVD, Walt Disney Home Video, 2002.
21 Quoted in Iwerks and Kenworthy, *Hand Behind the Mouse*, 62.
22 Garity, "Production of Animated Cartoons," 311–12.

storyboard in close collaboration with the sound and music departments. Only after the score and sound effects were settled would animators begin executing the final drawings, ensuring synchronization. In the early days, in addition to matching "noises" with their utterance or points of impact, the music was as percussive as possible to emphasize "sync points."[23]

Compare *The Opry House* (1929, dir. Walt Disney and Ub Iwerks),[24] the fifth Mickey Mouse (the third conceived with sound), to *Bright Lights*, the Oswald the Lucky Rabbit cartoon that had been released as a silent film one year before. Both are set in popular theatres and highlight vaudeville (music hall) type performances. *Bright Lights* marshals the full lexigraphic arsenal of silent cartoons. Toponyms set the scene and provide identifications: the theatre's flashing marquee reads "Vodvil." Signage steers the action: "To Stage Entrance," "Stage Entrance," and "Positively no admittance." A poster introduces "Mlle. Zulu Shimmy Queen/Admission 50¢." An empty animal cage is posted "Danger Keep Away" and another, "Lions Beware." In the final scene, the building is labelled "Theatre" and everyone flees through the "Exit." Just to avoid all ambiguity, paint cans are labelled "Paint." The animators avail themselves of sight lines to show Oswald's gaze when he spies the 50¢ price, not that there was any doubt about what he was looking at. Standard hieroglyphs appear throughout: The lucky rabbit's heart beats in his chest, then self-animates over his head; impacts produce star emoji; music emitters pop up, ranging from Oswald's whistles to the full orchestra's blasts, all accompanied by proper note emoji. Swirling bubbles and emotion lines show a dizzy doorman's state of mind.

Of course, cartoons like *Bright Lights* were not 'silent,' in the sense that, like all pre-talkies, live music probably accompanied them.[25] Certain scenes of *Bright Lights* appear to have anticipated locally supplied generic or improvised music for the opening vaudeville toponyms, the shots containing note emoji, and the stage numbers, which might have triggered appropriate hoochie-coochie and acrobatic performance themes – but this is conjectural. Whatever live musical accompaniment that existed would have depended on the music director/conductor, and would have varied among theatrical venues, even between performances. Recorded sound, thanks to exposure sheets, "canned" those myriad improvisational interpretations, substituting a single permanent music-image performance.

23 Jacobs, *Film Rhythm After Sound*, 58–108.

24 *Mickey Mouse in Black and White, Volume Two*, Disc 1, DVD, Walt Disney Home Video, 2004.

25 Goldmark, "Before Willie," 227.

The Opry House has vestiges of silent hieroglyphic conventions. It uses signage toponyms to create the animated world: "Opry [Opera] House," "Big Vaudville [sic] Show," a "Yankee Doodle Girls" poster, a window marked "Tickets," and an "Asbestos" curtain. There are still impact stars and shock lines when a hippo falls onto Mickey and during his piano recital.

There are, however, striking differences between the new and old films. The emoji from the silents have diminished. When the *Opry House* orchestra plays, we hear its actual music, so no note emoji are required. When the cats' tails are pulled and they emulate a bell ringing act, we hear their musical meows, but neither notes nor words amplify them. Mickey marching and whistling the "Yankee Doodle March" yields no flying notes. A mischievous kitten pours soap into an instrument. Bubbles emerge in just the right size and sequence so that, when the kitty pricks them, they play a measure from Georges Bizet's *Carmen*. We *hear* each note synched to individual bubbles, but there are no *visible* emoji. Speech and sound onomatopoeia, like "Ow!" and "Pop!" are conspicuously absent.

Except for a few very brief caesurae between 'acts,' the musical track is continuous, and so is the action synchronized to it. One comes away feeling that music animates this cartoon world, where sounds produce actions, not vice versa. The main interest in Mlle. Zulu's silent dance is erotic and visual. Seemingly a caricature of Josephine Baker, she shakes her coconut brassiere and grass skirt as her shimmying black body twists 'rubber hose' style. An attraction in *The Opry House* is Mickey – in drag – impossibly contorting his body to sway with the 'Egyptian' music, oscillating his big white breasts and popping off his belly button. The shift to klezmer music[26] in the finale summons Mickey's Hebrew alter ego and he transforms into a stereotypical caricature that reminds us of many vaudeville entertainers' Jewish roots. Unlike the effects in *Bright Lights* that call for a linear reading of the story and its integral written texts, *The Opry House* asks us to marvel at Mickey's malleability, so aptly identified by Sergei Eisenstein, as *plasmatic*.[27] The soundtrack invites us to immerse ourselves in Mickey's acoustic world, defined and dictated by melomania, where physics, causality, and body integrity bend to the inescapable fascination of music.

In September 1928, while the sound recording sessions for the first Mickeys were in progress, Disney and his animator Iwerks greenlit musician Carl

26 Identified as "Khosn, Lale Mazl Tov" ("Congratulations, Bride and Groom") by Sigmund Mogulesko. Goldmark, *Tunes For 'Toons*, 32.

27 Plasmaticness: "rejection of once and forever allotted form, freedom from ossification, the ability to dynamically assume any form." Eisenstein, *Eisenstein on Disney*, 21.

Stalling's idea for a "musical novelty" film that would not feature the Mouse.[28] The result, of course, was *The Skeleton Dance* (1929, dir. Disney),[29] which became the template for the Silly Symphonies – as well as the Looney Tunes, Merrie Melodies and other melomanic series that followed in the 1930s. As the name implies, the original concept was for each film to present pastiches of light-classical and popular pieces (as long as they were in the public domain), accompanied by humorous animated performances. The 'symphonic' animation glided, bopped, and throbbed beneath the synchronized syncopation of the score.

By mid-1929, Disney had decided that both the Mickey Mouse series and the Silly Symphony cartoons "will be set to a definite [rhythm] and we will have no more straight action to a mere musical background."[30] Relying at first on strict beats to the bar, Disney established the primacy of musical tonality, syncopation, instrumentation, and structure.

The third Symphony to be completed was *Springtime* (1929, dir. Iwerks).[31] Like most of the early ones, it is truly melomanic, as Merritt and Kaufman describe it: "The smooth, even strokes of the cello are picked up in rippling lines of water, while the violin's mad coloratura, playing counterpoint to the cello, gets picked up in the dancing squiggles of the frog's reflection."[32] Admired at the time for foregrounding music and precisely synchronizing image and action, students today tend to be put off by the relentlessly repeating horizontal movements that become monotonous. Hence the impression of musical madness.

A closer look at *Springtime*, though, highlights the distance this new animation form had travelled since the lexigraphic silent days. Instead of a written toponym, the film begins with a "carnivalesque opening shot"[33] establishing the scene with multiple layers of action – throbbing cattails, a burbling waterfall, and floating butterflies – all keeping time with "Morning" from Edvard Grieg's *Peer Gynt*. The dancing vignettes succeed each other to the tunes of "Whispering Flowers" (Franz von Blon), "Dance of the Hours" (Amilcare Ponchielli), and "Gaîté Parisienne" (Jacques Offenbach).[34]

28 Merritt, Kaufman, *Walt Disney's Silly Symphonies*, 5.

29 *Walt Disney Treasures: Silly Symphonies – The Historic Musical Animated Classics*, DVD, Walt Disney Home Video, 2001.

30 Disney, letter to Charles Giegerich, July 26, 1929, quoted in Barrier, *Animated Man*, 70.

31 *Walt Disney Treasures: More Silly Symphonies, Volume Two*, DVD, Walt Disney Home Video, 2006.

32 Merritt and Kaufman, *Walt Disney's Silly Symphonies*, 9.

33 Crafton, *Shadow of a Mouse*, 203–6.

34 Merritt and Kaufman, *Walt Disney's Silly Symphonies*, 60.

The animators used standard production techniques they had innovated in the silent period to extend their labour, but now they have pressed them into service as rhythm creators. These included background scene cycles to create an illusion that the subjects were moving across a broad vista. Individual drawing cycles matched the beats of the music. Retracing drawings created multiplied dancing figures. These were done by retracing one drawing several times. The new hatched chicks, for example, began with one drawing that was traced to make two chicks, then that drawing was flipped and traced to make another pair facing the opposite direction. The result was a chorus of chicks dancing in two symmetrical pairs, their beaks, feet, and winglets tapping out the music – all derived from a single drawing. Similarly, the frog's drawings were flipped vertically to sync his dance to his reflection in the water (a moment Gilbert Seldes singled out as "a grace-note of wit over the broad humorous symphony of the whole picture"). Except for a few croaks, the only sounds in this world are music and natural sounds mimicked in music. Seldes claimed that the Symphonies "are true sound pictures, without dialogue. [...] In them the animals and vegetation, purely incidental to Mickey Mouse, are brought into the foreground, and go through a wide range of activity – dances, skating contests, and so on – to the accompaniment of a reiterated musical theme."[35]

The Silly Symphonies shared a trait with films like *Bright Lights* and *The Opry House* – their performances were patterned on the types of live entertainment one might experience in a vaudeville theatre, an 'opry' house – or in 'encapsulated' live-action film equivalents of the sort that Vitaphone was presenting.[36] Disney's cartoon stage shows with dancing skeletons, animals, and 'vegetation' displaying themselves frontally to the audience and moving in choreographed lateral movements (except for a few startling aberrations when they swooped into the foreground) connected with audiences accustomed to watching similar live spectacles.

The cartoon-vaudeville connection was vividly demonstrated in *The Opry House* finale, when Mickey performs a 'trick piano' number. Bowing to the audience, he embarks on a well-synchronized piano recital with his gloved fingers playing the keys more or less accurately, until he shows off and falls off the stool. When the piano begins playing itself, Mickey goes on the attack, pounding the keys with his fists – until the instrument gets uppity and even the stool joins in the rebellion (Fig. 4.1). After a pitched battle, followed by reconciliation, all three principals take their curtain call.

35 Seldes, "Mickey-Mouse Maker," 26–27.

36 Crafton, *The Talkies*, 268.

Figure 4.1: *The Opry House* (1929, dir. Walt Disney and Ub Iwerks).

Mickey's tour de force presentation is an explicit media convergence. His routine parodies a standard vaudeville staple, pianist versus piano. These comedians 'animated' their pianos by tormenting and teasing them, until the instruments revolted. Herb Williams (1884–1936), who performed the same act with his "miraculous piano" for twenty-five years, is a template for Mickey's bit.[37] Vitaphone filmed the comedian in a short film, appropriately titled *Vaudeville* (1934, dir. Joseph Henabery).[38] Williams announces that he will attempt "a piano solo while standing on my head." Among the piano's pranks, the keys chatter on their own, the legs fall off, and it drenches the maestro in beer. It is out to get him! Like Mickey's, Williams' break-away piano has a will of its own and strikes back when stricken (Fig. 4.2). The inanimate object refuses to yield to the agency of the artist, whether Mickey or Williams. As adapted for animation, cartoonists no doubt relished bringing the anthropomorphized instrument to life and giving it a will of its own. Mickey, the animated being, is stymied when he animates the seemingly inanimate piano, which turns out to be sentient and already self-animated.

37 "Obituary of Herb Williams," *New York Times*, October 2, 1936, 52.

38 *Vitaphone Cavalcade of Musical Comedy: Shorts Collection*, DVD, Disc 4, Warner Bros. Archive Collection, 2009.

Figure 4.2: Herb Williams in *Vaudeville* (1934, dir. Joseph Henabery).

Many Mickey Mouse and Silly Symphony stories mimicked variety and music hall programmes with their song-and-dance revues, stage-bound compositions, screwball comedians, and familiar routines. When re-performed in cartoons, they become rhetorical figures of *metalepsis*, identified by Erwin Feyersinger as the "paradoxical amalgamation of ontologically distinct worlds."[39] Films that re-performed theatrical numbers derived much of their energy from confounding competing levels of visual and aural exposition. Many viewers in 1929 would have enjoyed *The Opry House*, if not inside an actual opera house, then in a similar familiar space dedicated to variety performances. Mickey's stage was a metaleptic version of the stage over which the film was being screened. The frieze-like dancing scenes in the Silly Symphonies were patterned on the front-of-curtain acts to which theatre patrons were accustomed, with the action confined to the tight downstage area while the next act assembled behind the curtain. Sometimes, this was also the function of a short film – to play on a screen dropped in front of the curtain to entertain and distract the audience during changes of acts.

The uncouth barnyard animals that comprise the audience in *The Opry House* are wildly appreciative, oblivious to the mistakes and musical transgressions in the show they are watching. Does this scene metaleptically

39 Feyersinger, "Diegetic Short Circuits," 279.

represent Disney's attitude toward the cinema audiences who were clamouring indiscriminately for the novelty of "all singing, all taking, all dancing," regardless of quality?

The big question fans were asking was if and when Mickey would talk, that is, enunciate audible speech. Aside from a few inarticulate squeaks, he did not speak right from the start. It was not until mid-1929 in *The Karnival Kid* (1929, dir. Disney and Iwerks)[40] that we hear him sing and talk – briefly. As a carny hot-dog vendor, he puts his trained wieners through their paces. In *The Haunted House* (1929, dir. Disney and Jack King),[41] Mickey's "Mammy!" channels his inner Al Jolson, in an obvious homage to *The Jazz Singer* (1927, dir. Alan Crosland). Both these initial forays into speech are highly reflexive. Mickey, the animated character in *The Karnival Kid*, is also Disney's avatar. Their roles have them 'directing' their subjects – hot dogs and the animation staff – to 'speak.' In *The Haunted House*, a grim ghostly figure demands that Mickey play the organ, mirroring the pressure from audiences and competing studios to have Mickey and his animated cohort sing and talk.

Conclusion

During the silent era, animators and audiences enjoyed cartoon performances, in part, as reading for meaning. While related to pre-existing popular graphic traditions, animation hieroglyphs often were embellishments signifying something that was already obvious. From the producers' viewpoint, they were shortcuts that alleviated the expensive labour-intensity of the animation process. While filmmakers assimilated the tradition, they also used the special opportunity afforded by moving pictures to expand the vocabulary beyond the capabilities of comics. They set their emoji in motion, metamorphosed them into props, body parts, and surprising juxtapositions.

It's commonplace to characterize the animation industry of the silent years as spewing out mediocrity, and their cartoons as 'filler.'[42] Another way to think of it, though, is that the system in the 1920s was conducive to a fairly stable set of production and artistic practices that prioritized

40 *Walt Disney Treasures: Mickey Mouse in Black and White*, DVD, Walt Disney Home Video, 2002.

41 *Walt Disney Treasures: Mickey Mouse in Black and White, Volume Two*, DVD, Walt Disney Home Video, 2004.

42 "[...] in 1928 – new cartoon series were simply not very attractive to most distributors. Cartoons' brief burst of popularity in the early 1920s was long past; of the major distributors, only Paramount and Universal now offered cartoons to theaters" (Barrier, *Animated Man*, 58).

a defined range of expression, the lexigraphic mode of production and reception. Animators relied on uniform conventions and iconography, honed for humour and narrative efficiency. The films prized character comedy (especially Chaplinesque), slapstick, and visual outlandishness, while adhering to 'classic' cinematic storytelling structures, rather than valuing unfettered artistic innovation. Occasional avant-gardism, as seen for example in the wild graphics of *Felix Dines and Pines* (1927, dir. Pat Sullivan [Otto Messmer])[43] were tolerated, in this case, by rationalizing the distortions as hallucinations. Conventions adapted from comic strips were fine-tuned by cartoonists, but I want to be clear: animation did not somehow evolve from comic art. Film borrowed heavily from many cultural prototypes and adapted them freely for its own needs.

The coming of sound profoundly affected the performance mode of animation. When sound became an option, was the situation like Chion's colourful description of the transition? "When, with the talkies, real sound entered the screen, acting like the late arrival who thinks the party can't start without him or her, people realized that the cinema had, in fact, gotten along perfectly well without it (except for speech)."[44] This formerly revisionist view of the talkies, that the coming of sound caused a disturbance in the classical mode of production that 'corrected' itself after a few years by assimilating many conventions established pre-sound, now has been widely accepted. But this brief survey suggests that this schema for mainstream cinema did not apply to animation. It seems that the older 'standard story' of the talking pictures' disruptiveness remains a viable explanation. In animated cartoons, the switch to sync sound and melomania really did cause a rapid and irrevocable change in styles, modes of production, and reception.

Cartoons' newfound sound strategy coincided with the populace's general enthusiasm about new audio-visual convergences. Sound pushed producers away from the lexigraphic mode related to comics, and toward live performance concepts, as mediated through radio and filmed vaudeville-style acts, such as those produced by Vitaphone. The recorded soundtracks made permanent the serendipitous music that had come from orchestra pits. The visuals in the Mickey Mouse and Silly Symphonies capitalized on prevailing entertainment expectations, particularly the frontal displays and friezelike side-to-side movements of the popular music stage.

Based on the critical and popular reception of Disney's shorts, moviegoers apparently welcomed the change. Critics did not mourn the good old days

43 *TS-39 Felix the Cat Vol. 5*, DVD, Tommy Stathes Cartoons on Film, 2008.
44 Chion, *Film, a Sound Art*, 16.

of cartoon silents. There is no evidence that anyone was disappointed when closely timed music brought these animated worlds to life, or when stars like Mickey finally did 'speak.'

This period of the 'tyranny of the metronome,' as the animators came to regard it, was short-lived. Simplistic mickey mousing declined in the mid-1930s. According to Jacobs, "The aim was no longer to create a simple match between movement and music but to integrate pacing, the timing of the animation, and the score. The synchronization between movement and music became more selective so that important moments could be accented and phrases of movement defined between prominent sync points."[45] But Disney never returned to his lexigraphic roots. Instead, he rapidly discarded the comic strip-influenced templates of his former cartoon cohort, like the Fleischers, Sullivan and Messmer, and Paul Terry, not to mention his own silent output, and alternatively looked to producers of live-action sound comedies, like Hal Roach, for his new mini-movie aspirational model.[46]

Thanks to Daniel Goldmark for helpful comments and identification of music elements.

Bibliography

Barrier, Michael. *The Animated Man: A Life of Walt Disney*. Berkeley: University of California Press, 2007.

Barrier, Michael. *Hollywood Cartoons: American Animation in Its Golden Age*. New York: Oxford University Press, 1999.

Barrier, Michael, Milton Gray, and Bob Clampett. "Interview with Wilfred Jackson (1973)." *MichaelBarrier.com*, July 31, 2015. Accessed February 2, 2023. http://www.michaelbarrier.com/Interviews/Jackson1973/Jackson1973.html.

Borge, Victor. "Phonetic Punctuation." Accessed February 2, 2023. https://www.youtube.com/watch?v=Qf_TDuhk3No.

Chion, Michel. *Film, a Sound Art*. Translated by Claudia Gorbman. New York: Columbia University Press. 2009.

Conley, Tom. *Film Hieroglyphs: Ruptures in Classical Cinema*. Minneapolis: University of Minnesota Press, 2006.

45 Jacobs, *Film Rhythm After Sound*, 107.

46 Letter from Walt Disney to Roy Disney and Ub Iwerks, September 7, 1928; quoted in Barrier, *Animated Man*, 61.

Crafton, Donald. *Emile Cohl, Caricature, and Film*. Princeton, NJ: Princeton University Press, 1990.

Crafton, Donald. *Shadow of a Mouse: Performance, Belief, and World-Making in Animation*. Berkeley: University of California Press, 2013.

Crafton, Donald. *The Talkies: American Cinema's Transition to Sound, 1926–1931*. Berkeley: University of California Press, 1997.

Danesi, Marcel. *The Semiotics of Emoji: The Rise of Visual Language in the Age of the Internet*. London: Bloomsbury, 2016.

Eades, Mark. "First Draft of Mickey Mouse 'Steamboat Willie' Script on the Auction Block." *Orange County Register*, October 7, 2016. Accessed February 2, 2023. https://www.ocregister.com/2016/10/07/first-draft-of-mickey-mouse-steamboat-willie-script-on-the-auction-block/.

Eisenstein, Sergei. *Eisenstein on Disney*, edited by Jay Leyda. Translated by Alan Upchurch. New York: Methuen, 1988.

Eisner, Will. *Comics & Sequential Art*. Exp. ed. Tamarac, FL: Poorhouse Press. 2001.

Emojipedia. "Musical Note." Accessed May 30, 2023. https://emojipedia.org/musical-note/.

Feyersinger, Erwin. "Diegetic Short Circuits: Metalepsis in Animation," *Animation: An Interdisciplinary Journal* 5, no. 3 (2010–2011): 279–94.

Fischbach, Lisa. "Punctuation (1): The Linguistic Side." *Cast*, May 9, 2017. Accessed February 2, 2023. https://articles.c-a-s-t.com/punctuation-1-the-linguistic-side-9ba8eb11fbf2.

Garity, William. "The Production of Animated Cartoons," *Journal of the Society of Motion Picture Engineers* 29, no. 4 (1933): 309–22.

Goldmark, Daniel. "Before Willie: Reconsidering Music and the Animated Cartoon of the 1920s." In *Beyond the Soundtrack: Representing Music in Cinema*, edited by Daniel Goldmark, Lawrence Kramer, and Richard D. Leppert, 225–45. Berkeley: University of California Press, 2007.

Goldmark, Daniel. *Tunes For 'Toons: Music and the Hollywood Cartoon*. Berkeley: University of California Press, 2005.

Iwerks, Leslie, and John Kenworthy. *The Hand Behind the Mouse*. New York: Disney Editions, 2001.

Jacobs, Lea. *Film Rhythm After Sound: Technology, Music, and Performance*. Berkeley: University of California Press, 2015.

Kane, Brian. *Sound Unseen: Acousmatic Sound in Theory and Practice*. New York: Oxford University Press, 2014.

McCloud, Scott. *Understanding Comics: The Invisible Art*. New York: HarperCollins, 1994.

Merritt, Russell, and J. B. Kaufman. *Walt Disney's Silly Symphonies: A Companion to the Classic Cartoon Series*. Revised and Updated. Glendale, CA: Disney Editions, 2016.

Newsom, Jon. "'A Sound Idea': Music for Animated Films," *The Quarterly Journal of the Library of Congress* 37, no. 3/4 (Summer/Fall 1980): 279–309.
Seldes, Gilbert. "Mickey-Mouse Maker," *The New Yorker*, December 19, 1931, 23–27.

About the Author

Donald Crafton is a retired Professor of Film, Television, and Theatre at University of Notre Dame. He has received the Distinguished Career Award from the Society for Cinema and Media Studies, and the Jean Mitry prize for individuals or institutions distinguished for their contribution to the reclamation and appreciation of silent cinema. In 2001, the Academy of Motion Picture Arts and Sciences named him an inaugural Academy Film Scholar. He is the author of *Before Mickey: The Animated Film, 1898–1928* (1982), *Emile Cohl, Caricature, and Film* (1990), *The Talkies: American Cinema's Transition to Sound, 1926–1931* (1999) and *Shadow of a Mouse: Performance, Belief, and World-Making in Animation* (2013). He currently is writing plays and screenplays.

II

From Theory to Practice

5. "To Select, To Organize, To Sharpen"

Rouben Mamoulian, Sound Film Theory, and *Applause*

Michael Slowik

Abstract: Film scholars have praised Rouben Mamoulian's *Applause* for its early innovative use of sound. Less attention, however, has been paid to two key questions: precisely *how* unusual were such techniques when the film was released, and what *factors* encouraged Mamoulian to offer such an unorthodox sound approach? This article answers these questions by examining Mamoulian's beliefs about artistry in general, and film sound in particular, and by comparing *Applause* to the sound practices of other films at the time. Mamoulian's devotion to medium specificity and stylization drove him to conceptualize film sound as an element to be selected, manipulated, and harnessed to the story, an attitude that was unusual at the time but would soon become the Hollywood norm.

Keywords: medium specificity, stylization, foreground / background sound, dialogue

Rouben Mamoulian's *Applause* (October 1929) has been widely heralded as a landmark in the development of synchronized sound film practices. Sonically, scholars have pointed to its daring work in such areas as two-track recording, audio density, off-screen sound, and sonic expressivity.[1] Less attention, however, has been paid to two key questions: precisely *how* unusual were such techniques when the film was released, and what *factors* predisposed Mamoulian to break the ground he did?

1 The most in-depth exploration remains Fischer, "*Applause*." See also Barrios, *A Song in the Dark*, 315–16; Crafton, *The Talkies*, 334–38; Koszarski, "The Greatest Film Paramount Ever Made" and Milne, *Mamoulian*, 17–28.

Wiegand, D. (ed.), *Aesthetics of Early Sound Film: Media Change around 1930*.
Taylor & Francis Group, 2023
DOI 10.5117/9789463727372_CH05

Such questions are vital for grasping both the early history of sound film aesthetics in the United States and the career trajectory of Mamoulian, one of the most understudied directors in American entertainment. Though many of *Applause*'s stylistic methods would anticipate later sound film practices, we do not always recognize just how much of an anomaly *Applause* was upon its release in 1929. And while Mamoulian has been praised as an innovator, less attention has been paid to the extent to which *Applause* was the product of Mamoulian's broader, unified theory of art, and the role that Mamoulian's theatrical background played in his direction of the film. Far from a one-off experiment, *Applause* reflected the concerted effort of a director-artist to define sound cinema's essential properties and display ways that sound could be manipulated for narratively expressive purposes. Examining Mamoulian's theories and their realization in *Applause* helps uncover the methodology of a neglected director and enables us to see how filmmakers began to separate from stage-bound practices.

Filmmaking was brand new to Mamoulian when he directed *Applause* in 1929, but he benefitted tremendously from having spent years thinking rigorously about art in general, and theatre in particular. An Armenian born in Russia-controlled Georgia, Mamoulian enjoyed a childhood in which theatre was always nearby, as his mother performed in a theatre troupe throughout much of his youth. Sometime in the mid 1910s, Mamoulian attended Moscow University and studied briefly at the Moscow Art Theatre. Following a period that historians have found difficult to fully trace, Mamoulian directed his first play, *The Beating on the Door*, in London in 1922. Mamoulian then emigrated to the United States, first to direct opera and live movie prologues in Rochester, New York, from 1923 to 1926, and then to direct plays for the Theatre Guild in New York City. Mamoulian directed six plays prior to his debut film, *Applause*, but the most significant was his first, *Porgy* (1927), a critical darling and commercial hit that made him a major name on Broadway.[2]

Thanks especially to his constant work and freedom to experiment in Rochester, Mamoulian developed a theory of art in the 1920s – which he expounded upon in articles, interviews, and lectures – that includes concepts we would now categorize in terms of medium specificity. For Mamoulian, an artist – no matter the form that he or she worked in – was best served by *locating* the elements that enabled an art form to be called by its name, and

2 See Spergel, *Reinventing Reality*, 9–36, for the most thorough account of Mamoulian's years in Europe.

stylishly exploring the potential of those elements to express ideas about the subject matter. During his Rochester years, Mamoulian concluded that on the stage, it was *movement*, sometimes orchestrated with rhythm in mind, that defined theatre as a distinct artform and constituted theatre's most impactful feature. For Mamoulian, other elements (music, scenery, playscripts, flesh-and-blood actors) could be jettisoned and still result in something recognizable as theatre, but movement could not. Mamoulian thus saw movement as the lifeblood of the form, and consequently the domain within which stylization would be most beneficial.[3]

Where did Mamoulian's theory of art emerge from? Mamoulian, in interviews, always implied that it sprang full blown from his own head. By Mamoulian's account, his brief study with Yevgeny Vakhtangov at the Moscow Art Theatre (presumably at the First Studio, since Vakhtangov taught there at the time) focused heavily on naturalism, leading Mamoulian to mimic this approach in *The Beating on the Door*.[4] It was this use of naturalism – which Mamoulian discovered gave him no satisfaction when he watched the finished play – that prodded him to develop an opposing theory of stylization.[5] Though such an account squares with existing archival evidence, it is plausible that Mamoulian underplayed the extent to which his time in Moscow also *helped* him develop his theory of stylization. Joseph Horowitz raises the prospect that Mamoulian's later devotion to stylization stemmed from his encounters with Vakhtangov, including Vakhtangov's concept of "fantastic realism."[6] This is possible, though there is also evidence to suggest that Vakhtangov was still under the sway of naturalism in the mid 1910s.[7] Konstantin Stanislavsky also emerges as a possible influence, partly because of his sheer prominence in Russia at the time. Although Mamoulian apparently never studied with him, Stanislavsky did create the First Studio in 1912, and his presence loomed over it. Moreover, though Stanislavsky is commonly associated with realism, he could be quite versatile. In 1909, for instance, Stanislavsky started the Studio Theatre on Povarskaya Street with Vsevolod Meyerhold to experiment with music and symbolism, a form that – like Mamoulian's later stage and screen work – favoured stylization

3 Mamoulian, "The Essence of Theatrical Art." For another analysis of Mamoulian's theatre writings during his Rochester years, see Horowitz, "*On My Way*," 79.

4 Spergel, *Reinventing Reality*, 10.

5 Mamoulian, interview by Becvar, Tape 1, 6.

6 Horowitz, "*On My Way*," 10–3.

7 Vahktangov, for instance, was reportedly amazed by what Stanislavsky did with *Twelfth Night*, a 1917 First Studio production that broke the fourth wall. My thanks to Sharon Carnicke for this information.

over realism.[8] Stanislavsky's Moscow Art Theatre troupe also apparently once stayed at Mamoulian's home when he was a boy, an experience that Mamoulian acknowledged "was subconsciously ingrained in me."[9]

Whatever the influences, what mattered most is that when Paramount executives hired Mamoulian to work on films at their Astoria studio in early 1929, Mamoulian brought along his theories of medium specificity and stylization to the nascent domain of sound cinema. The timing was fortuitous for Mamoulian, because while it was just becoming clear that sound film would be permanent, practitioners had not yet worked out what a sound film would look and sound like. Mamoulian took film seriously as an art form, so rather than treat cinema as a mere extension of the stage, he instead sought to find and exploit attributes specific to sound film. As Mamoulian would explain in articles and lectures throughout the ensuing decade, this meant – first and foremost – that he would exploit the camera's ability to move and frame objects in particular ways, both of which Mamoulian felt were central to cinema and had no corollary on the stage.[10] But Mamoulian was also interested in how a filmmaker could manipulate *sound* in a medium-specific way that would express key narrative ideas.

Mamoulian's theoretical approach to sound can best be gleaned from two position papers he wrote in the late 1930s that advocated stylization over realism. The first paper, titled "Visual Anti-Realism," is a philosophical treatise on artistry and the reasons why, in Mamoulian's view, an artist should favour stylization. In this paper, Mamoulian contrasted the work of a scientist, who aims to state facts in an impersonal manner, with the artist, whose job is to transform facts by creating new values that are personally meaningful to the artist.[11] Mamoulian's second paper – titled "The Psychology

8 For considerations of Mamoulian's possible intellectual debt to Vakhtangov and Stanislavsky, see Spergel, *Reinventing Reality*, 33–41, and Horowitz, "*On My Way*," 10–12, 81. For a discussion of Stanislavsky and symbolism, see Carnicke, "Rethinking 'Stanislavskian' Directing," 99–104. More generally, I am indebted to Carnicke for helping me think through Mamoulian's Russian influences.

9 Horowitz, "*On My Way*," 11.

10 For two early statements on the subject, see Mamoulian, "The Use and Abuse of Perambulation," and Mamoulian, "Common Sense and Camera Angles." Mamoulian's theories bear resemblances to writers publishing in Russian, French, or German at the time, including Rudolf Arnheim, Béla Balázs, the French Impressionists, writers for *Close Up*, and the Russian montage theorists. Mamoulian knew these languages and may have encountered some of these writings. Still, Mamoulian's philosophies of film are so similar to his prior ideas about theatre that it may have been primarily his own established methodology that drove him to define cinema in the ways he did.

11 Mamoulian, "Visual Anti-Realism."

of Sound" – used this framework to consider an artist's role with respect to film sound. If an artist, Mamoulian argued, seeks to transform rather than reproduce, the artist's job is "to select, to organize, [and] to sharpen" any element that can be used to express ideas.[12] To determine how this general philosophy applied to film sound, Mamoulian drew a contrast between human ears and a microphone. Ears, Mamoulian asserted, are highly critical and selective instruments: they decide what to attend to versus what to merely hear. The microphone, in contrast, picks up sounds indiscriminately. Thus, since the microphone is an impersonal instrument more akin to science, a true film artist must transform the microphone into a critical device akin to the human ear by consciously *selecting* the components of the soundtrack for the audience member.[13]

Mamoulian's accuracy here can be questioned. Microphones *are* selective instruments – they record only certain sound waves – and by the late 1930s, when Mamoulian wrote the article, the rise of directional microphones had made them more selective still. But Mamoulian was not aiming at a theory of technology. Rather, his point was that microphones (outside of their recording capabilities) make no decisions; they collect rather than organize. To Mamoulian, it was thus incumbent upon the film artist to perform the essential tasks of selection and manipulation. "No matter what the *natural* sounds of a scene may be," Mamoulian argued in "The Psychology of Sound," "it is up to [film artists] to select them – to subdue, eliminate, exaggerate, and otherwise *distort* these sounds, *in complete defiance* of Nature."[14] Any noise that one takes for granted in life, Mamoulian argued, should not be present in film at all. Each sound should instead be selected with rigorous care to convey the emotional implication and mood of the scene. In short, film sound should not be a casual or accidental capturing of reality. Rather, as in art in general, "everything should be done according to a consciously controlled artistic and dynamic pattern."[15] As with stylization in general, Mamoulian felt that for film sound, "the more completely and successfully an artist battles realism, the more real will be the result to the audience spectator."[16]

It would be difficult to overstate how strongly at odds with the mainstream this philosophy would have been when Mamoulian made *Applause* in

12 Mamoulian, "The Psychology of Sound," 2.
13 Ibid., 3–4.
14 Ibid., 4 (emphases in the original).
15 Ibid., 11.
16 Ibid., 11.

1929. Based on scholarship, and my own review of every available all-talking film from 1928 and 1929, it is clear that many films open with precisely what Mamoulian rails against: the presentation of a wide range of noises that showcases the technology's ability to capture and reproduce sounds regardless of their expressive place in the story. Such stretches are often narrative-less and centred on the sounds of a particular space, from the din of Tin Pan Alley (*The Broadway Melody*, February 1929, dir. Harry Beaumont) or the New York City streets (*Sunny Side Up*, October 1929, dir. David Butler), to the western frontier (*The Virginian*, November 1929, dir. Victor Fleming). Many films also dwell on the presentation of accents and stuttering, even when irrelevant to the plot. Song plugging – the repetition of a song to drum up sheet music sales – often took priority over narrative coherence.[17] Donald Crafton has pointed to the 1928–1929 season, in particular, as insistently featuring a "'you can't miss this!' approach to sound."[18] In many other cases, movies take on a filmed-play quality, with dialogue-heavy scenes featuring only the types and qualities of sound that would be heard on the stage. Film sound in the earliest years often seemed less about being integrated expressively into a formal system of meaning, and more about offering noises as discrete, unaltered attractions.

Mamoulian wrote "The Psychology of Sound" in 1938, when filmmaking had drifted away from such approaches, and one could argue that his essay reflected this shift. Mamoulian's sound work in *Applause*, however, hews closely to the principles he would later lay out, and in "The Psychology of Sound," he pulled extensively from *Applause* to illustrate his arguments. In *Applause*, rather than presenting sound as if it were the labour of a field-worker who documents existing sounds, a showman who presents them as discreet attractions, or a stage director, Mamoulian wished to aggressively announce film sound's unique ability to be selected, manipulated, and harnessed to storytelling ends.

Applause and Foreground Sound

We can best assess how *Applause* accomplished these aims by distinguishing between foreground and background sound. Of the two, foreground sound appears to have been less manipulated on the whole. One crucial scene,

17 For a thorough account of popular song's dominant role in the early sound era, see Spring, *Saying It with Songs*.
18 Crafton, *The Talkies*, 271.

however, deviates sharply from this tendency: a five-minute scene – captured in a single take – in which burlesque star Kitty (Helen Morgan) tries to soothe her sheltered seventeen-year-old daughter, April (Joan Peers), to sleep after April witnesses Kitty performing in a degrading burlesque number. Near the end of the scene, Kitty gently sings to April a lullaby version of the song she performed at the burlesque house – "Give Your Little Baby Lots of Lovin'" (by Joseph A. Burke and Dolly Morse) – while April, who has been raised in a convent, simultaneously murmurs a prayer to help herself fall asleep. The idea of overlapping Kitty and April's voices during this scene was almost certainly Mamoulian's – it appears not in the typed script but in Mamoulian's handwritten annotation in his copy of the screenplay.[19] In addition to overlapping their voices, Mamoulian wanted them at the same volume level, as he felt this would better articulate the larger clash between the burlesque and convent worlds that the film dramatizes.[20]

Overlapping voices was not, by itself, new when Mamoulian shot *Applause*, but his desire for voices at equal volume posed a logistical challenge. Ordinarily, a 1929 sound technician tasked with miking two physically close speaking characters would hang a single microphone above the actors. Such a setup, however, would have caused Kitty's singing to overwhelm April's murmured prayer. The eventual solution appears to have differed somewhat from what Mamoulian would later claim, but it still constitutes a striking act of manipulation for the period. According to Mamoulian, two microphones were used: one near Kitty for her singing, and one under April's pillow as she prays.[21] These two microphones recorded onto two different tracks, which were then rerecorded together onto a single strip in post-production, with the volume of April's prayer raised to the same volume level as Kitty's song.[22]

In interviews, Mamoulian implied that both microphones were rigged on the set, meaning that Kitty and April's voices were both recorded live.[23] The omnidirectional nature of 1929 microphones, however, likely made such an

19 Annotation in October 8, 1929 *Applause* screenplay, *Applause* file, RMP.

20 Mamoulian, "The Art of Films," 27.

21 Mamoulian, interview by Hare, 127.

22 Mamoulian offers a detailed description in Mamoulian, Columbia University oral history, 79–80.

23 In earlier interviews, Mamoulian did not specify when he recorded Kitty's track (Mamoulian, "The Art of Films," 26–27; Mamoulian, Columbia University oral history, 79–81). In later interviews, however, he implied that Kitty and April were recorded simultaneously (Mamoulian, *The American Film Institute Seminar* (1981), 1TA/P4–P10; Mamoulian, interview by Gallagher and Amoruco, 16–17).

arrangement impossible.[24] Kitty sits very close to April for much of the scene (Fig. 5.1), including while April prays, making it unlikely that microphones could have isolated Kitty or April's voices. Multiple tracks *do* appear to have been used, however, as one can detect an increase in ground noise as April prays and a simultaneous reduction in the volume level of Kitty's singing. Moreover, at least one track was surely recorded live, since the clanks of the moving camera bungalow can be heard on a few occasions during this long-take scene. It thus seems most likely that April's prayer and much of Kitty's spoken dialogue were recorded on the set into a single microphone, while most of Kitty's singing was recorded either beforehand or afterward. The camera's close proximity to April as she prays necessitates tight lip synchronization that would be difficult unless miked live, and at one point in the conversation, Morgan accidentally talks over Peers and must repeat her line, which could have occurred only if both actresses were recorded live at that point with a single microphone. For her humming and singing, however, Kitty is frequently off camera, and when she is visible, her humming does not require lip synchronization. If Kitty's humming/singing was, indeed, recorded at a different time, this opens up the possibility that the beginning of the scene – which features Kitty singing in near darkness across the room from April and being interrupted by April – also used rerecorded vocal tracks.

Whatever the precise method, the most significant aspect was *not* – as Mamoulian would later assert – the fact that he used two microphones to record two different voices or the sheer use of rerecording, which can be found in prior films.[25] The scene's originality instead stems from Mamoulian's decision to rerecord *dialogue* tracks and adjust their volume levels in postproduction. As Lea Jacobs has demonstrated, studios in the early sound era shied away from rerecording dialogue because they feared that the rerecording process would ruin intelligibility.[26] Filmmakers also generally avoided raising the volume level of an already recorded track in postproduction because doing so often noticeably increased ground and surface noise.[27]

24 I am grateful to Lea Jacobs – and also Martin Barnier – for helping me work through the technical logistics of this scene.

25 Mamoulian, Columbia University oral history, 80–82. Early 1929 releases like *In Old Arizona* (January, dir. Irving Cummings) and *Alibi* (April, dir. Roland West) feature characters who speak too far away from each other for a single omnidirectional microphone to catch their words, making it quite likely that multiple microphones were used. For information on early rerecording, see Jacobs, "The Innovation of Re-recording" and Slowik, "Experiments in Early Sound Film Music."

26 Jacobs, "The Innovation of Re-Recording," 12.

27 Ibid., 10–11.

Figure 5.1: In *Applause* (1929, dir. Rouben Mamoulian), Kitty's and April's microphone levels are adjusted – and their voices are likely recorded at different times – to express the film's tension between burlesque and convent life.

The decision to record voices on separate tracks may seem like a mere technical matter, but it has broad ramifications for a filmmaker's role with respect to recorded sound. In June of 1929, if a filmmaker called for two microphones to record dialogue, the expectation would have been that both would be mixed onto the same track during shooting, with technicians determining the microphones' volume levels ahead of time. Yet on-set sound mixing lacked precision. Once the information from two microphones is rerecorded onto a single track, a filmmaker cannot separate back out sound from the two microphones to further adjust volume levels.[28] Mamoulian instead opted for the method that would give him the most control in post-production, and therein lies a key assumption about film sound's potential. *Through two-channel recording, Mamoulian was conceptualizing directly recorded sound – including dialogue – as an object for filmmaker manipulation.* As commonsensical as this notion may seem in filmmaking today, in 1929 – and indeed, for a number of years afterward – filmmakers regularly prioritized (and even showcased) the apparatuses' ability to

28 Richard Koszarski points out this alternative recording method in "The Greatest Film Paramount Ever Made," 439.

faithfully reproduce the sound record of a given space. Jacobs notes that as late as 1932, the industry felt strongly that the volume levels of on-set dialogue shouldn't be tampered with in post-production, and not until the mid-1930s did the rerecording of dialogue become commonplace.[29]

Also crucial to this recording method is Mamoulian's prioritization of what he saw as the unique properties of sound cinema. Part of what appealed to Mamoulian about raising April's voice to Kitty's level was that he *couldn't* do it in a medium like the theatre, where Kitty's onstage singing would have drowned out April's near whisper.[30] Moreover, rerecording itself, as Jacobs has pointed out, was in the late 1920s unique to cinema.[31] The scene thus stands as an early testament to Mamoulian's belief that sound should be manipulated in cinema-specific ways to enhance vital aspects of the story.

Applause and Background Sound

If the lullaby scene constitutes a one-off moment for manipulated foreground sound in *Applause*, Mamoulian's efforts to select, isolate, and expressively manipulate background sound are far more pervasive. The consistency with which Mamoulian treats such sound as a spatial and expressive tool is all the more remarkable because in the late 1920s, background sound appears to have been among the most unsettled of aesthetic questions. American films in 1928 and 1929 display a dizzying – and often inconsistent – array of decisions surrounding the basic question of whether to provide background sound at all. Depending on the film – or even particular shots within a single scene – one finds everything from the inclusion of apparently every background sound to a complete absence of background noise, even when sound-producing elements appear clearly in the background of a shot. Though seemingly every conceivable use of background sound can be found, probably the most frequent one is what Rick Altman has called the "on/off switch approach," in which some element – an open versus closed door, an on-screen versus off-screen producer of sound – dictates the presence or absence of background sound. A closed door might eliminate loud music entirely – as it does repeatedly in *The Lights of New York* (July 1928, dir. Bryan Foy) – or nearby sounds might be heard only when visible in the frame, as

29 Jacobs, "The Innovation of Re-Recording," 12.

30 Mamoulian discusses the medium-specific nature of his sound approach to the lullaby scene in Mamoulian, "The Art of Films," 25–27.

31 Jacobs, "The Innovation of Re-Recording," 6.

is frequently the case in *The Big Trail* (October 1930, dir. Raoul Walsh).[32] Even in these instances, however, inconsistency remains the norm, with rules seemingly established in one shot or scene quickly jettisoned in the following shot or scene.

In *Applause*, Mamoulian's rule for handling background sound *is* fairly consistent, and it falls under what we might call perceptual realism, a fairly uncommon approach in the late 1920s. In general, Mamoulian provides background sounds whenever they could plausibly be heard, regardless of their onscreen visibility. Moreover, as Lucy Fischer has pointed out, he consistently adjusts volume and reverberation levels in accordance with the distance between camera and sound source (an approach that may have also made use of rerecording).[33] Though such an approach might, at first blush, appear naturalistic rather than expressive, it gave Mamoulian the opportunity to 1) *select* sounds that would be especially expressive, and 2) *manipulate* these sounds if their distance from the camera might plausibly result in sonic distortion.

Applause's background burlesque music offers a useful illustration. *Applause*'s numerous scenes set inside burlesque houses feature the constant sounds of the burlesque performances, including scenes taking place in the wings of the stage and the theatre's dressing rooms. Such music's prevalence, alone, conveys the sense that burlesque indelibly moulds every facet of the central characters' lives (a key idea in the film),[34] but Mamoulian also uses the reverberant nature of these spaces to transform burlesque music into grotesque noise. Early in *Applause*, when the chorus line files into Kitty's dressing room to visit Kitty and the newborn April, Mamoulian offers an onstage rendition of Irving Berlin's "Turkey Trot" that is so reverberant that the singer sounds as if she is shouting or even barking the lyrics. It is this extraordinarily 'bad,' ear-ringing music that hangs over April and Kitty in this scene, and it establishes in vivid terms their relationship to burlesque. Later, when seventeen-year-old April watches the conclusion of her mother's burlesque performance from backstage in dismay, Mamoulian again uses a high level of reverberation. Here, because the song – Art Fitch, Kay Fitch, and Bert Lowe's "Doin' the Raccoon" – is

32 Altman, "Inventing the Cinema Soundtrack," 351–52; Altman, "Establishing Sound," 28.

33 Fischer, *"Applause,"* 244–45. A 1929 trade-journal account stated that "sound perspective" was attained in "an important story sequence" by hanging microphones in two different parts of the theatre, recording them on separate tracks, and adjusting volume levels via post-production rerecording ("Sound Perspective Used in 'Applause,'" 10).

34 As Fischer puts it, "sounds from other locations aggressively pursue" the film's central characters (*"Applause"* 239).

performed so quickly and with a balance level favouring the instrumental accompaniment, the reverberation warps the performance into a jumbled mess of unintelligible and nearly tuneless singing. In both scenes, sonic distortion helps present burlesque as an overbearing presence for the female performers who toil within it. By consistently allowing off-screen music to bleed into backstage shots of Kitty and April, and by taking advantage of the microphone's ability to capture a single level of reverberation for all the film's spectators, Mamoulian uses background sound and distortion as expressive tools at a time when most filmmakers were grappling simply with whether to use such sound at all.

Mamoulian's location sound likewise demonstrates his embrace of cinematic sound devices and his insistence upon using even sound recorded in less-controlled environments to express narrative ideas. *Applause* was one of the first films to use directly recorded sound in urban space, and because of this, Mamoulian would have felt considerable pressure to simply showcase the technology's ability to capture and present these sounds. Yet these sounds, too, are selected, controlled, and harnessed to the ideas of the story. For instance, Mamoulian's personal screenplay annotations indicate that he was responsible for a chorus singing Franz Schubert's "Ave Maria" on the soundtrack as April departs the Wisconsin convent. Upon a downbeat of this song, Mamoulian abruptly cuts to the loud and abrasive sound of steam escaping as April's train comes to a stop at Pennsylvania Station in New York City and Mamoulian's on-location sound work begins. "Sound of trains," Mamoulian wrote into the margin of his screenplay at this moment.[35] On the soundtrack, a constant muffled roar – likely attained by on-location sound recording at Penn Station – accompanies Mamoulian's hidden-camera footage of April inside the station. Once April steps outside the station and hails a taxi, we hear a barrage of honking car horns, squealing brakes, and roaring motors all the way to Kitty's hotel. Through this carefully curated soundtrack of aural contrasts between the convent and New York City, the audience hears New York City as April does: a startling, confusing, and dangerous space.

Sharply contrasting with these harsh, punctual noises is the background sound Mamoulian selects for April's first two dates with Tony (Henry Wadsworth) on the Brooklyn Bridge and the top of the AT&T building. Here, Mamoulian uses height to justify a far quieter backdrop, with only the roar of a passing airplane in the AT&T building scene intruding upon this space. Even this, in fact, was no accident. In the film, the airplane soars overhead just after April accepts Tony's proposal of marriage, and Mamoulian hired

35 Annotation in October 8, 1929 *Applause* screenplay, *Applause* file, RMP.

the plane's pilot to fly low in the sky past the lovers, apparently breaking a city law in the process. Mamoulian would later see this moment as a prime example of how a filmmaker could select and emphasize certain sounds to punctuate and accent the emotion inherent in the scene – for him, the sound of the airplane was pivotal for expressing the joy of the moment beyond what any actor could convey.[36]

Even the film's most technically challenging location scene – April and Tony's apparent break up at a subway stop late in the film – features Mamoulian orchestrating narratively expressive sound. Shooting on location at New York City's Chambers Street subway stop and using only direct sound,[37] Mamoulian – in a single take – begins with virtually no background sound, a quietude then enhanced by the loud arrival and departure of the train, a sonic "monster" that appears to whisk Tony away forever. After the train departs with Tony aboard, April weeps, and environmental sound becomes all but non-existent once more, an occurrence justified by the fact that the train has departed, but one that also underscores the tragedy of the moment. Even when shooting on location, Mamoulian consistently configured environmental sound as an object to be selected and timed precisely with the narrative beats of the story.

Intercut with April and Tony's breakup is a scene, shot indoors at the studio, that features Mamoulian at arguably his most sonically manipulative. Kitty, having poisoned herself so as to free April and Tony to marry without guilt, sits alone in her darkened apartment grimly waiting for death. To express Kitty's mindset, Mamoulian composed an off-screen diegetic noise symphony featuring the sounds of car engines, a ticking clock, trolley bells, traffic-cop whistles, car horns, the voices of a nearby couple on an elevator, and an ambulance siren. These sounds grow louder and denser, and when they reach their most cacophonous level, Kitty, now teetering on the brink of insanity, shouts, "April!" and rushes out of the apartment.

The specific sounds used in this scene were, according to Mamoulian, carefully planned and layered. The sounds were produced,

> [...] in a separate sound room in the studio with about eight property men handling different horns, klaxons, drums and roller skates, while I stood by on a chair conducting this conglomeration, giving [cues], indicating entrances and building up a gradual crescendo. These traffic sounds

36 Mamoulian, "The Psychology of Sound," 8–9.

37 Koszarski, "The Greatest Film Paramount Ever Made," 440.

> started low and soft, then kept growing in volume. At the end, the sound of a siren was heard, presumably approaching from a distance, then passing by Helen Morgan's hotel and then, as if tangled up in the traffic, shrieking with a loud urgency and despair.[38]

Through it all, as Mamoulian pointed out, Morgan remains rooted to her chair in a darkened space. Mamoulian was drawn to the idea that "the sound montage was acting for her [...] the clamor of that dramatic, organized traffic, expressed the rise of her emotions. When the siren reached its highest pitch and volume, it was like the agonized cry of Helen Morgan's heart – she jumped up crying out the name of her child 'April!'"[39] Such a cacophony arguably echoes a similar city-noise barrage in *Sunrise* (September 1927, dir. F. W. Murnau), which featured synchronized music and sound effects only.[40] In the context of an all-talking film, however, Mamoulian's scene was rare at the time – if not unique – for featuring such an elaborate construction of environmental sound for narrative and psychological purposes.

None of this is to say that film sound had a single identity or purpose for an innovator like Mamoulian to 'discover.' Confusion abounded in the late 1920s, but the period was also ripe with promising experiments that simply did not endure, and these are worth our attention. *Mamoulian's* efforts, however, often *did* anticipate subsequent sound filmmaking, and such a story is equally important. *Applause* stands as one of the earliest sound films to overtly signal Hollywood's future direction: soundtracks whose constructed and selective nature aided and expressed narrative ideas. Mamoulian's interest in medium specificity and stylization would, in subsequent films, lead to pioneering work in such domains as the diegetic voice over (*City Streets,* April 1931) and synthetic sound (*Dr. Jekyll and Mr. Hyde*, December 1931). For the industry as a whole, it would become commonplace by the end of the 1930s to rerecord dialogue tracks and manipulate them in postproduction, while background sound became sparser, more selective, and more narratively relevant. Mamoulian's theoretical framework, and its manifestation in *Applause*, is a vital – and still largely untold – story in the history of sound film aesthetics.

38 Mamoulian, "The Art of Films," 33.

39 Ibid., 34.

40 My thanks to Martin Barnier for pointing to this precursor.

Bibliography

RMP = Rouben Mamoulian Papers, The Library of Congress, Washington, D.C.

Altman, Rick, McGraw Jones, and Sonia Tatroe. "Establishing Sound." *CiNéMAS* 24, no. 1 (2013): 19–33.

Altman, Rick, McGraw Jones, and Sonia Tatroe. "Inventing the Cinema Soundtrack: Hollywood's Multiplane Sound System." In *Music and Cinema*, edited by James Buhler, Caryl Flinn, and David Neumeyer, 339–59. Hanover: Wesleyan University Press, 2000.

Barrios, Richard. *A Song in the Dark: The Birth of the Musical Film*. New York: Oxford University Press, 1995.

Carnicke, Sharon Marie. "Rethinking 'Stanislavskian' Directing." In *The Great European Stage Directors, Volume 1: Antoine, Stanislavski, Saint Denis*, edited by Peta Tait, 91–111. London: Methuen Drama, 2019.

Crafton, Donald. *The Talkies: American Cinema's Transition to Sound, 1926–1931*. Berkeley: University of California Press, 1997.

Fischer, Lucy. "*Applause*: The Visual and Acoustic Landscape." In *Film Sound: Theory and Practice*, edited by Elisabeth Weis, and John Belton, 232–46. New York: Columbia University Press, 1985.

Horowitz, Joseph. *"On My Way": The Untold Story of Rouben Mamoulian, George Gershwin, and Porgy and Bess*. New York: W. W. Norton & Company, 2013.

Jacobs, Lea. "The Innovation of Re-Recording in the Hollywood Studios." *Film History* 24, vol. 1 (2012): 5–34.

Koszarski, Richard. "The Greatest Film Paramount Ever Made." *Film History* 15, vol. 4 (2003): 436–43.

Mamoulian, Rouben. *The American Film Institute Seminar*. May 27, 1981, RMP.

Mamoulian, Rouben. "The Art of Films." Lecture given to "History of the Motion Picture" class at Columbia University, Museum of Modern Art, December 6, 1939, RMP.

Mamoulian, Rouben. "Common Sense and Camera Angles." *The American Cinematographer* 12, vol. 10 (1932): 8–9, 26.

Mamoulian, Rouben. "The Essence of Theatrical Art." Lecture delivered in Rochester (NY) 1923–1924, RMP.

Mamoulian, Rouben. "An Interview with Rouben Mamoulian." Interview by John A. Gallagher, and Marino A. Amoruco. *The Velvet Light Trap*, no. 19 (1982): 16–22.

Mamoulian, Rouben. "Interview with Rouben Mamoulian." Unpublished interview by William Becvar. March 20–22, 1973, RMP.

Mamoulian, Rouben. Oral History Transcript, Columbia University, New York, December 1958, RMP.

Mamoulian, Rouben. "The Psychology of Sound: The Second in a Series of Papers Challenging the Realistic Conception of Art." July 1938, RMP.

Mamoulian, Rouben. "Rouben Mamoulian: An Exclusive Interview." Interview by William Hare. In *American Classic Screen Interviews*, edited by John C. Tibbetts and James M. Welsh. Lanham, MD: Scarecrow Press, 2010.

Mamoulian, Rouben. "The Use and Abuse of Perambulation." Lecture given to the American Society of Cinematographers, July 19, 1932, RMP.

Mamoulian, Rouben. "Visual Anti-Realism: The First in a Series of Papers Challenging the Realistic Conception of Art." July 1938, RMP.

Milne, Tom. *Mamoulian*. London: Thames & Hudson, 1969.

Slowik, Michael. "Experiments in Early Sound Film Music: Strategies and Rerecording, 1928–1930." *American Music* 31, vol. 4 (2013): 450–74.

"Sound Perspective Used in 'Applause.'" *Film Daily*, September 15, 1929, 10.

Spergel, Mark. *Reinventing Reality: The Art and Life of Rouben Mamoulian*. Metuchen, NJ: The Scarecrow Press, Inc., 1993.

Spring, Katherine. *Saying It with Songs: Popular Music and the Coming of Sound to Hollywood*

Cinema. New York: Oxford University Press, 2013.

About the Author

Michael Slowik is Associate Professor of Film Studies at Wesleyan University. He is the author of *After the Silents: Hollywood Film Music in the Early Sound Era, 1926–1934* (2014) and numerous articles, many of which pertain to Hollywood's early sound era, for example "Diegetic Withdrawal and Other Worlds: Film Music Strategies before *King Kong*, 1927–1933." His latest book is *Defining Cinema: Rouben Mamoulian and Hollywood Film Style, 1929–1957*, forthcoming from Oxford University Press.

6. Futurists and 'Homogenizers' in Early Soviet Sound Film

Oksana Bulgakowa

Abstract: This chapter discusses a theoretical concept by Nikolai Anoshchenko, who attempted to reframe cinematic speech by approaching it not as a verbal but as a sonic phenomenon. The semantic power of sound was supposed to become greater than that of dialogue or monologue – an idea that shows a close relation to the sound poetry of Russian Futurists. The idea of mobilizing the phonosemantics of sound was only partly put in practice but the first few projects of the Moscow and Leningrad studios offer especially interesting case studies in this regard. These films suggest a homogenization technique that affects both voice and noise and share similar strategies of manipulating acoustic phenomena.

Keywords: film theory, Russian Futurism, film music, speech in early sound film, phonosemantics

The sound revolution took over ten years in the Soviet Union: until 1938, films were released in both sound and silent versions. Early Soviet sound film is known mostly for the manifesto "Statement on Sound" (1928) by Sergei Eisenstein, Vsevolod Pudovkin, and Grigori Alexandrov, in which the authors advocated for asynchronous sound, for a discord between sound and image, in order to recombine the two in new ways.[1] These ideas were partially realized in Dziga Vertov's *Entuziasm* (*Enthusiasm*, 1931) and Pudovkin's *Dezertir* (*The Deserter*, 1933),[2] and they were included in Eisenstein's several unrealized projects (the sound version of *Staroye i novoye/The Old and the*

1 Eisenstein/Pudovkin/Alexandrov, "Statement on Sound."
2 Thompson, "Early Sound Counterpoint."

Wiegand, D. (ed.), *Aesthetics of Early Sound Film: Media Change around 1930*.
Taylor & Francis Group, 2023
DOI 10.5117/9789463727372_CH06

New, 1929; *Sutter's Gold* and *An American Tragedy*, both 1930; and *MMM*, 1933). In these scripts, Eisenstein developed the idea of an inner monologue based on the conflict between an inner and an outer voice, between the figurative and the abstract, the visual and the acoustic, the objective and the subjective, attributed alternately to the image and the sound. Vertov's film and Eisenstein's ideas are the most extensively analysed examples, even if the latter were rarely examined as acoustic phenomena and products of a sonic imagination. Until recently, however, other concepts and works have rarely been explored.[3]

The early stage of sound film in the Soviet Union was shaped by passionate debates, experiments, and a variety of approaches. Technicians and governmental institutions discussed whether it would be better to import American sound equipment or to invest into domestic sound-recording technology, and which of the two competing labs – Pavel Tager and Alexander Shorin – should be given priority.[4] Shorin's system was prioritized by the state-run studios in Moscow, Leningrad, and Kiev. Tager's system was used at Mezhrabpomfil'm, a joint venture between the German section of the Workers International Relief (WIR) and the Rus' studio. The first sound film by Mezhrabpomfil'm, *Putevka v zhizn'* (*Road to Life*, 1931, dir. Nikolai Ekk), was very successful, but the original soundtrack of the film was reconstructed, that is, destroyed twice – in 1958 and 1977. During this process, the dialogue was recorded again and post-synchronized, along with new music and new mixing.[5]

The avant-garde filmmakers who were conscious of their brands in silent movies – montage instead of a linear plot, masses instead of individuals, types instead of individualized characters – tried to transfer their principles to sound film. Pudovkin tested the sounds of the masses, while Vertov returned to the Futurist experiments of his youth with the montage of urban and industrial noises. The Futurist heritage was also revived in experiments with artificial, synthetic sound based on a combination of graphic design (arches, dots, sine waves, squares, serpentine lines) and manipulation of the optical soundtrack, comparable with Oskar Fischinger's experiments from 1932.[6]

The influence of the Russian Futurists can also be traced in another aspect of early sound film that mobilized the semantics of sound expressed

3 An exception is Kaganovsky, *The Voice of Technology*.
4 Posener, "To Overtake and Outstrip Hollywood."
5 Bulgakowa, *Golos kak kul'turnyi fenomen*, 511–22.
6 Smirnov, *Sound in Z*, 175–238.

by the voice. My contribution analyzes how the first Soviet sound films used 'wordless' voices as a self-sufficient means of expression by treating the voice as a purely sonic phenomenon, as in *Dela i liudi* (*Men and Jobs*, 1932, dir. Alexander Macheret), produced in Moscow. Then, I turn to the first three productions of the Leningrad studio – *Odna* (*Alone*, 1931, dir. Grigori Kozintsev, Leonid Trauberg), *Zlatye gory* (*Golden Mountains*, 1931/1936, dir. Sergei Yutkevich), and *Vstrechnyi* (*Counterplan*, 1932, dir. Friedrich Ermler, Sergei Yutkevich).[7] They offer especially interesting case studies of early Soviet sound aesthetics, suggesting a homogenizing technique that affected both voices and noises, and which would define the use of sound in Soviet films for the next decade.

The Semantics of Sound

The rise of sound film established a new kind of filmmaking that was supposed to structure sonic chaos into meaningful expression, which would, in turn, instruct the viewer who did not yet know how to listen or what to hear. The filmmakers would now, quite literally, guide the viewers' ears as much as their eyes. Listening was no longer to be a merely physiological process but an exercise in semantic prowess: audiences would not just hear but extract meaning from all acoustic phenomena (music, noises, and voices).

By deploying the paralinguistic qualities of the human voice – intonation, volume, and rhythm – as a means for conveying meaning, cinema once again aspired to become established as an international art form that could surmount all language barriers. The most consistent formulation of this theory comes from Nikolai Anoshchenko,[8] in a report on sound film delivered at the May 11, 1929 meeting of the Moscow chapter of the Association of the Workers of Revolutionary Cinematography (ARRK). Anoshchenko proposed to classify cinematic sound into noises, "sounds of performing objects" (e.g., trains, automobiles), and what he called "differentiated sounds" (*differentsirovannye zvuki*), which included music and the human voice

7 All these films are part-talkies: they have 'silent' episodes, montage sequences, and synchronized dialogues, but in varying proportions. *Odna* was started as a silent film and post-synchronized. *Zlatye gory* was re-released in 1936 in a re-edited 2,605-metre version (instead of 3,585 metres), but it is not known what happened with the original soundtrack. *Vstrechnyi* was started as a full talkie but has some 'silent' passages.

8 Anoshchenko (1894–1974) graduated from the Moscow Film School in 1927 and was sent to Germany to study film technique in 1929. He did not make any sound films himself but developed patents in film technology.

(where sounds are 'differentiated' according to distinct tonal or phonetic values). Yet, according to Anoshchenko, speech in the new sound cinema should be used in a very limited manner. It would need to be merged with music, which would in turn arise from "an immediate sensory emotional source – screaming, moaning, threatening, [...] only those vocal expressions that are ripe with emotional content," and only then, only eventually from an ideological cause: "the voice will sound as if it were music, and in this sense its sound will be self-sufficient for expressing meaning."[9]

This would become possible because the differentiating values in the sound of the human voice were easy to detect immediately, since "spoken language is characterized by vocal modulations in frequency and tempo – that is, the same applications of sound as those employed in music. Intonation can be understood more easily than words. Voice matters more than discourse."[10] Anoshchenko suggests that aural expressivity should be the main consideration when writing film dialogue. Combinations of melodic vowels with sibilant and fricative consonants ought to give rise to high-energy sonic forms with their own connotations and semantic field. This requires researching and determining specific combinations of sounds that produce such sonic forms and then identifying the means of intensifying them. There are low vowels – "ooh," "oh," "ah," "eeh" – and high vowels, devoiced and voiceless consonants, sounds that are either soft or taut like a coil waiting to be sprung: "We have to study and analyse sound creatively, we have to research *dyr byl shchil.*"[11] Here, Anoshchenko directly invokes the expressive ideas of the Russian Futurists by citing the earliest and most famous Zaum poem by Aleksei Kruchenykh, performed in December 1912. Zaum was the transrational language developed by Russian Futurist poets, who created unusual, evocative sound combinations that were indeterminate in meaning; the aim was to create a new language free from grammar and syntax, organized according to phonosemantics and rhythm, and capable of articulating a wordless, affective sound-speech, "simple as a cow's mooing" (Vladimir Mayakovsky). Futurist poet Velimir Khlebnikov hoped to find the essential meaning of word roots in their sounds and thus create a new, magically universal language, like the "language of the birds" or "language of the gods."[12] In the 1920s, these practices were analysed by the Russian

9 RGALI, *fond* 2494, *opis'* 1, *edinitsa* 222, list 2–3.
10 Ibid., list 3.
11 Ibid., list 12.
12 Janecek, *Zaum*, 137–38.

formalists, and they resonated in the work of the Institute of the Living Word directed by Sergei Bernshtein.[13]

The earliest theorists of Soviet sound film took up many of these ideas, and some directors tried to apply them in their work.

The Voice of the Masses vs. the Inner Voice

Vsevolod Pudovkin edited the soundtrack for his sound film *Dezertir* separately from the film footage, guided by the same rhythmic principles he used to edit his silent films. He worked on this film for almost three years. *Dezertir* consists of about 3,000 shots, as was common for a silent picture. To Pudovkin, the dialogue, the screaming of a single word (in the rally scene), and the noises (the symphony of the sirens at the beginning of the film) had the same semantic value. The actors' speech highlighted one word in each sentence, and Pudovkin saw in this technique a free and powerful means of expression, which avoided the "conversational" style.[14]

> [I] cut the sound as freely as the image. I used three distinctly different elements: the speech, the sound close-ups of interruptions – words and fragments of sentences spoken in the crowd – and the total of the crowd with its hubbub of varying intensity, recorded independently of the image. [...] Each sound was cut separately, and the images associated with it were sometimes much shorter than the soundtrack, sometimes equal in length. [...] Sometimes, with the help of scissors, I inserted the hubbub of the crowd into a sentence. I found out that by juxtaposing various sounds I can create a clearly defined, almost musical rhythm, a rhythm that is produced and amplified from one frame to another, growing like waves, until it reaches the climax of the emotional impact.[15]

Pudovkin's strategy was different from that of Eisenstein, who tried to convey the psyche of a protagonist who had lost language under the influence of an affect. At those moments, an inner monologue would emerge, the only form of spoken word that should be used in sound film according to Eisenstein. Eisenstein did not envision inner monologue as a simple narrative

13 Olenina, *Psychomotor Aesthetics*, 1–40.

14 Pudovkin, *Sobraniye sochinenii*, 155.

15 Ibid., 165–66.

voice-over. Instead, he strove to imitate the rhythmic, asyntactic flow of words, sounds, and images that is characteristic of thought. He interpreted visual and acoustic elements as an expression of the antagonism between consciousness and the unconscious. The visual and the acoustic would change roles: either picture or sound could represent the conscious at a given time, while the other element would, in turn, remain symbolic or abstract. Eisenstein wanted this second, non-representational element to be pure affect – like a black screen or an asemantic sound. In the script for "An American Tragedy," he reduced speech to one word charged with affect in a special dramaturgy:

> He [Clyde] sits down on the couch again, nervous and shivering, he picks up the paper he had thrown away and rereads the article. And while he is rereading it with wide-open eyes, the whisper from afar gradually creeps up till it forms the word: "KILL!" In a strange, gradual way the phrase spoken by the whisper forms and forms until at last it pronounces and repeats the whole word: "KILL! KILL!" And from this moment the action begins to work alone: the line of the thoughts of a distracted man, leaping from one fact to another, suddenly stopping – departing from sane logic, distorting the real rumor between things and sounds ... They are likewise distorted, and a whisper becomes the whistle of a storm, and the storm cries out "Kill," or the whistle of the storm becomes the movement of the street, the wheels of a streetcar, the cries of a crowd, the horns of motorcars, and all beat out the word: "Kill! Kill!" And the street noises become the roar of the factory machines, and the machines also roar out "Kill! Kill!"
>
> Or the roar of the machines descends to a low whisper, and it whispers again: "Kill! Kill!" And at this moment a pleasant, unemotional voice slowly reads the newspaper article: *Fifteen years ago, a similar accident occurred, but the body of the man was never found.*[16]

The voice (integrated into a rich soundscape) was understood as a pure, ecstatic expression, bursting out of the unconscious, from an idée fixe, into a monosyllabic exclamation ("Kill!"). This word was supposed to become a leitmotif throughout the film, embodied in three shapes – as image, voice, and intertitles, which correspond to Eisenstein's three stages of mental representations (the image as a sensuous analogy of the visible; sound,

16 Quoted in Montagu, *With Eisenstein in Hollywood*, 286–87.

which is still sensuous but more abstract; and finally, the pure abstraction of a written word). Eisenstein analysed a similar practice with reference to Khlebnikov's poem *Spell of Laughter* (*Zaklyatiye smekhom*) from 1910, in which the poet achieves the effect of trance with constant variations of word formations based on the same root.[17]

Pudovkin's approach, unlike Eisenstein's, was not based on psychological aspects; instead, his editing only followed rhythmic considerations. Because dialogues often take time, sometimes several minutes, they were difficult to fit into Pudovkin's film style with its predominance of brief close-ups. Instead of dialogue, he thus relied on short acoustic units, such as single words or even just syllables with their phonosemantic associations. This practice, typical of Pudovkin, was inspired by the Futurists' experiments with sound repetitions.

In 1929, Adrian Piotrovsky, head of the script department at the Leningrad studio and a gifted translator of Aristophanes and Sophocles, wrote:

> Obviously, we will more and more often use noises and all sorts of non-harmonized sounds, which are more expressive and unusual for the ear of listeners. Music, songs and human speech should be put in second place, speech should happen most likely in the form of emotional yelps, sharp intonations, and by no means in the form of semantic dialogues.[18]

In 1938, Piotrovsky would have to vacate his job, before he was arrested and executed.

It was possible to preserve the famous Russian montage style in the talkies, but did the character types promoted by the avant-garde directors align with a vocal culture? Working-class and rural performers rarely possessed voices whose sound was sufficiently expressive, and sometimes their voices were even perceived as pathological, as the poet Alexander Blok suggests in his diary: "Our servant. A peasant girl. I suddenly […] heard her voice. Something terrible, […] a nasal voice from her toothless mouth. […] Degeneration-inducing horror because connected with something unknown."[19] The voice betrayed too much of this unknown, and the types disappeared from the Soviet screen; by the mid-1930s, they were fully replaced by theatre actors with well-trained voices.

17 Eisenstein, *Metod/Die Methode*, 395–409.
18 Piotrovsky, *Teatr: Kino: Zhizn'*, 230.
19 Blok, *Sobraniye sochinenii*, 144.

The Mechanic and the Organic: What Is Nature?

However, these well-trained voices were transformed into electric phantoms on the screen: they were screeching, buzzing, thundering. Still, in the Soviet film discourse, the unnatural qualities of recorded voices went largely unnoticed, which explains the absence of several key topics that dominated the discussion in the West from the debate (phonogenic vs. non-phonogenic voices, sonic realism, the media-specific characteristics of electrical voices, etc.), even though the technologies did not differ all that much. The Soviets used German microphones and amplifiers as well as US-made soundboards for mixing. But the problem of phonogenic voices, which ended the career of some silent film stars, was discussed only in relation to Hollywood.[20] One explanation for this "deafness" of the Soviet filmmakers and theorists might be the cultural tradition shaped by Russian literature, which was less sensitive to the differences between 'natural' and 'artificial' voices. The traumatic stories about strange mechanical voices coming from singing machines were absent from the Russian debate.[21] Thus, even if the technologies used in Russian movies were the same as in the West, they did not change the conceptualization of the voice as a natural phenomenon and bearer of an ontological essence, a view that ignored the transformation of the natural voice on the screen into a highly stylized product of technology due to microphones, loudspeakers, equalizers, filters, etc.

As the earliest microphones recorded sound, they performed a kind of selection process: their ability, and frequent failure, to capture certain pitches, specific volume ranges, and inevitable changes in timbre led to various aberrations. A very high soprano, a low bass, or an intimate, velvety voice did not register on the early optical soundtrack printed on Soviet film stock and often sounded distorted. Both the low and the high frequencies were cut off, and the recording gave the voices an unpleasant timbre. Actors could not be positioned close to the microphone (this closeness also produced distortions), so immediacy and intimacy vanished: the voice always sounded like it was coming from a distance, and the speaker sounded like they were addressing the invisible human

20 See, for instance, the transcripts of a discussion at the ARRK on February 17, 1930, in which journalist Lev Nikulin describes the Hollywood sound practice (RGALI, fond 2494, opis' 1, edinitsa 303, list. 8).

21 Bulgakowa, *Golos kak kul'turnyi fenomen*, 69–103.

masses off-screen rather than their on-screen interlocutors. This effect of a 'public' sound was additionally reinforced by strong reverberation, which early Soviet sound films could not yet neutralize. At times, a simple line delivered during a close-up came off as a scream. One had to speak "slowly, very slowly, the slower the better, almost syllable by syllable, and emphasize every single word. [...] The vocal timbre was changed, and corrupted, in a very palpable way."[22] Microphones limited the selection of high voices, and sound-on-film recording gave them a light, metallic timbre. This stripping down of voices and noises – a kind of involuntary acoustic minimalism – led to a peculiar cultural dynamic of the voice in Soviet cinema.

The only solution to this problem was to use not just theatrical technique but also vocalization: to emphasize, as a professional singer would, sonorous consonants, lengthy vowels, and to err on the side of over-enunciation and a sluggish tempo. This operatic technique was preserved on the Soviet screen throughout the thirties. Technique tried to pass itself off as the natural, ideal voice, whose intonational dynamics were governed by the principles of highly conventional sonic organization – namely, music – not least because certain standard pitches of a tonal scale simply happened to be easier to record than many other sounds (especially various kinds of noises). From a technological point of view, this was due to the fact that musical sounds were more stable in their reproduction, in contrast to the uneven fluctuations of other sounds.[23]

It is not by chance that several early sound films featured a musician or a singer as the protagonist. But instead of backstage musicals, Soviet filmmakers turned to films about industrial production set in factories or on big construction sites. Nevertheless, all these films were defined by the music whose melodic power subdued noises.

Vertov would not have liked any of these films since they did not document authentic sounds but created them artificially in the studio. He wanted to "walk out of the muffled coffin of the sound studios" and record the "iron clanking and fearful roar of Donbas" on location.[24] But by varying the speed of his recordings, Vertov nevertheless changed the pitch of the sound and introduced gradation and differentiation.

22 Kalganov, "Pervye gody Leningradskogo radio," 206–7.

23 Helmholtz, *Die Lehre von den Tonempfindungen*, 15.

24 Taylor, Christie, *The Film Factory*, 302–3.

Dela i liudi (*Men and Jobs*)

Alexander Macheret's first sound film is a very expressive suggestion of how to deal with industrial noises, human voices, and music. The film thematizes the competition between the American and the Soviet production systems, embodied in a sonic dialogue. But Macheret dismissed words as bearers of meaning. His protagonist, a Russian foreman played by Nikolai Okhlopkov, struggles with words and fails to articulate a simple speech during an obligatory meeting. Okhlopkov delivers a virtuosic performance as a drunken worker trying to express himself in front of the mirror. His senseless soliloquy consists of broken sentences, repetitions, interjections, and pauses: "Now I … I … well, that is … uh … I … um … well … what I am saying …" This impaired speaker cannot communicate with an American engineer who came to teach the Russian workers how to deal with the American machines. For the Russian screen characters (and spectators), this American speaks an incomprehensible language, and his Russian assistant does not know how to translate Soviet neologisms, such as *udarnik* ("shock worker") or *sorevnovanie* ("socialist competition"), into English.

Macheret relied on the phonetic expressivity of intonation in the dialogues, or he used industrial sounds that replaced words. The latter were imitated by a special musical apparatus constructed for this film, which transformed them into distinct and comprehensible signs characterizing the machines' movements (as energetic, helpless, drunken, etc.), so that the anthropomorphized machines – instead of the protagonists – seem to speak and develop a dialogue. At the end of the film, having mastered the machine, Ochlopkov's worker can speak without stumbling, and the American engineer, who has been listening to his jazz record throughout the film, throws it away and opens the window (and his ears) to the new sonic landscape of Soviet society.[25]

The film is based on an inconspicuous substitution. The music here is mostly diegetic, coming from a gramophone record, a radio, an orchestra on stage, an accordion on- and off-screen. But since the industrial noises function musically as a substitute for the dialogues (or underscoring them apart from a few sentences), the film achieves a strange effect – an illusion of diegetic noises produced by invisible instruments. The nuanced rhythmic sounds adopt the traditional role of film music by creating accents, mood, and tension, by commenting on and accelerating the action, and using melodic intonation instead of speech. But Macheret's

25 Widdis, "Making Sense without Speech," 102–4.

approach of rejecting words as bearers of meaning was too ambiguous and counterproductive for Soviet film in the 1930s, which was looking for a clear, verbally articulated political message. The film received positive reviews[26] (unlike the devastating critiques of Vertov's industrial cacophony),[27] but its aesthetics would not be developed further (and was used only in comedies).

In contrast to this effort to discover a new acoustic dimension, the first sound films of the Leningrad studio – *Odna*, *Zlatye gory*, and *Vstrechnyi* – offer interesting case studies of a musical homogenization technique that affected both voices and noises and returned to traditional musical solutions. Macheret followed the same principle but 'in disguise,' presenting music and voices as 'noise.'

The Sound of Music

The musical scores for all three films from the Leningrad studio were composed by Dmitry Shostakovich, who collaborated very closely with the same team: sound designer Leo Arnstam and sound engineer Ilya Volk. The musical styles of the films are very different, but they share the same strategy of manipulating different kinds of acoustic phenomena. Music is the medium for translating one sonic dimension into another. Noise and often silence turn into music; conversational speech becomes singsongy.

In several scenes that take place in factory halls, the industrial background noises are actually produced by musical instruments. Arnstam described how he created them in the otherwise realistic *Vstrechnyi*, using a simple rhythmic pattern played by four basses, several clarinets, violins, flutes, and trumpets joined by the hum of the lighting equipment. As he explains, film critics falsely perceived these 'noises' as music (which flattered him), but when "we showed the 'noises' to the factory workers, they accepted them as authentic."[28] Only in some instances is the use of musical instruments obvious. In one lyrical sequence, the asphalting of the city is accompanied by a symphonic arrangement, with a xylophone played with wooden sticks, and when a worker (Vladimir Gardin) drinks and bites into a pickle, this action is ironically accompanied by a tuba that simulates his burping. In *Zlatye*

26 Tsekhanovskii, "Dela i liudi kak professional'nyi urok," 15–16.

27 Radek, "Dve fil'my," 4.

28 Quoted in Akhushkov, *Kak voznik i stavilsya Vstrechnyi*, 100.

gory, this kind of musical illustration takes over completely: gun shots are rendered by timpani, percussion is used to represent the clatter of hooves.

The most creative approach can be found in *Odna*. In the first part of the film, there is a scene in a porcelain shop that is supposed to sound like an 'opera of singing objects.' Kozintsev recalled that he sought to distort the sound and produce an effect as if porcelain cups and toys were singing. They did not succeed, and the sounds of the singing dishes were replaced by a traditional tenor. Even earlier in the film, 'voices' of objects (a kettle, a stove, a tram) can be attributed to musical instruments. A tram enters into a dialogue with a tuba, birds converse with a barrel organ, the clatter of a typewriter turns into a vocal recitative.[29] The most striking aspect is the way in which the film emphasizes the electric, ghostly qualities of the human voice: the disembodied voices from radios and loudspeakers intrude into and invade the world of human relations, control the fate of the heroine, and deprive her of her voice.

The plot of the film is simple, but the artistic execution is very refined. A young teacher played by Elena Kuzmina does not want to accept a job far from her Leningrad home, in a small village in Altay, but she eventually follows the state's order. The voice of the state comes from a three-piece street loudspeaker (like a three-headed dragon) or is heard as the emotionless voice of an invisible body. The soundscape of the village is created by the howls of a shaman, whose voice embodies the voice of a dead horse (the shaman was brought to Leningrad and recorded at the studio). Kuzmina's voice (during the whole film she only utters one short sentence, an outcry) is 'replaced' by two singing voices (male and female), then by a flute (during the lesson she soundlessly leads at an Altay school), and, at the end of the film, by a fanfare. In the finale, her fate is narrated on the radio by an unfamiliar male voice. The substitution of one voice with another, with a noise (e.g., of a typewriter), or a musical instrument creates a peculiar effect: several key conversations in the film sound like instances of musical combat (but not like the combat of noises heard in *Dela i liudi*).

Boris Alpers interpreted the plot of as if it anticipated Fritz Lang's *Das Testament des Dr. Mabuse* (*The Testament of Dr. Mabuse*, 1933), noting in his review that the film, built on a musical principle, 'silences' the heroine with electric voices (the radio, the telephone), which almost kill her at the end.[30] At the same time, Kuzmina's 'alien' voices intone a normative sound, which connotes a certain manner of articulating (and thinking). In the

29 Kozintsev, *Sobranie sochinenii*, 197.

30 Alpers, *Dnevnik kinokritika*, 104–7.

culturally diverse Soviet Union, cinema and radio 'levelled' the variety of different dialects and voices, which also meant that a young woman could easily sing as if she were a young man.

The fundamental principle of levelling voices or replacing them by noises or music is also central to the aesthetics of *Zlatye gory*. Here, all the actors speak slowly, with lengthy pauses between two words in the same sentence, which emphasizes the song-like quality of the intonation, and they stretch out certain vowels, such as "ah" and "oh." The screenwriter of *Zlatye gory*, Alexei Chaplygin, selected words defined by the four vowels "a," "o," "i," "e," which the actors stretch almost endlessly, and even the intertitles contain many words dominated by "a" and "o." There is an analogy here to a common practice in popular songwriting, with the lyrics chosen so that they contained the vowels that sounded best with a certain singer.

The scene of a work break lasts five minutes (as indicated by a diegetic sign), during which words and pauses are distributed rhythmically, as if with the help of a metronome. Dialogues are underscored by an invisible (probably nondiegetic) balalaika. A worker: "Zhena rozhaaaet/The wife is giving birth [pause of three seconds] ... Doooctora naaado/A physician is needed [pause of three seconds]. Doktor daaarom ne poidjooot/But he won't come for nothing [pause of four seconds] Plaatiiit'?/How can I pay?" And so on, until a messenger reports that God has sent the worker an heir ("nasleeednika"), and the announcement is followed by a dance. After a change in the melody, the scene is interrupted by the question: "A zhenaaa?/ And the wife?" During a twenty-second pause, the messenger slowly takes off his cap, the widower's hand freezes in the air, and the balalaika melody comes in again as the widower turns his back to the camera, no longer waiting for an answer. In this way, the music substitutes for the silence (a semantic silence, not the dead soundtrack with its deafening electric creaking used in other early sound films).

In this scene, the intonations of all characters follow the rising and lowering of the melody played by the balalaika, and the endings of the sentences coincide with the musical cadences. This type of musically stylized dialogue emphasizes a leitmotif of the film, which was conceived as an adaptation of a folk song ("Zlatye gory"). Shostakovich composed an elaborate symphonic score. This nondiegetic music accelerates the action and the montage and accompanies the most dramatic episodes. In the first part of the film, most dialogues are underscored, and when the underscoring disappears in the second part, the actors continue to speak as if singing the recitatives. Legato vowels are pronounced at a high pitch and used to neutralize the effect of explosive consonants by way of stretching and melodic movement on the vowels "a," "o," and "e." This manner of speaking resulted in a long duration of each individual word.

The same principle of vocal stretching and musical intonation was adopted in *Vstrechnyi*, where the directors refrained from musical underscoring for the most part, but the actor Vladimir Gardin developed an elaborate system of singsong with "emotional dominants," as he called them. These could only be expressed by vowels, which ensured a "correct sonic reaction" that "colored the emotion [...] and mattered much more than the words' literal meaning."[31] A short informational phrase ("I graduated from a college") was recorded four times, and in the chosen take, joy fills all stressed vowels – "o," "u," "a" – quite evenly, without emphasizing individual words.[32] The hero's intoxication as played by Gardin in a different scene was described as follows: "All vowels are stressed, hoarse with drunken anger."[33] In this shot, the actor chokes on a vowel, which he stops stretching to focus on the explosive consonant "x" instead.[34]

For the nearly three decades that followed, Soviet cinema relied with remarkable consistency on this musical idiom to "level" and standardize sound, to balance noises and voices, and to regularly substitute musical vocalization techniques for conversational speech and normative intonation patterns. The ratio of equalization to outright substitution was determined by the conventions of specific genres: romantic comedy, musical, or melodrama. Yet in Soviet cinema, in contrast to European and American films, the musicalizing practice extended far beyond the musical comedy to films as diverse as *Karl Brunner* (1936, dir. Alexej Masliukov, Mechislava Maievskaia) about the plight of German Communists; the collective farm film *Chudesnitsa* (*Miracle Woman*, 1936, dir. Alexander Medvedkin); the romantic comedy *Devushka speshit na svidanie* (*Late for a Date*, 1936, dir. Mikhail Verner), and the war epic *Padenie Berlina* (*The Fall of Berlin*, 1949, dir. Mikhail Chiaureli). The plots of some sound films told the story of the education of a simple hero, whose speaking style evolves from vulgar language to cultured speech. The talkies taught people how to speak, but the characters did not stop singing. The voice was mostly embedded into the musical score, and its melodies dictated the intonation and rhythm of speech. The music translated emotion into stylized and tempered expression. Instead of the voice producing the impression of music, the music substituted the voice.

31 Gardin, "Kak ia rabotal nad obrazom Babchenko," 97.

32 Ibid., 98–99.

33 Ibid., 100.

34 Ibid., 97.

The speeches in Eisenstein's *Ivan Groznyi* (*Ivan the Terrible*, 1944, 1946–1958) used (underscored) musical recitatives with long pauses, the melodic rise of the voice, clear articulation, and the deliberate treatment of the duration of each individual syllable. Mikhail Chekhov, a famous actor trained by Konstantin Stanislavsky, was unpleasantly struck by this practice. In a letter from May 31, 1945, he reproached Eisenstein for destroying the famous Russian affect with his artificial manner. Chekhov reacted to the new vocal norm established in Soviet cinema's more recent realistic films; Eisenstein condensed this technique into a baroque form, which horrified Chekhov with its "unnaturalness," a dead, mechanical rhythm with unjustified pauses between words, and added emphasis due to the slowdown. According to Chekhov, this heavy, metric speech "excluded the possibility of a change in pace," it paralyzes the energy, kills the tension and the silence.[35]

Films of the 1930s produced a special kind of sound and electric voice. Their piercing, high, metallic, and intense sound created the impression of an artificial acoustic world. Music served to smooth this impression. *Dela i liudi* transformed voices into musical noises, the productions of the Leningrad studio translated voices into music. Moving pictures had practised this from the very beginning, and in some sense, the Soviet solution returned to silent film on a new level.

Bibliography

Akhushkov, Shamil, ed. *Kak voznik i stavilsya Vstrechnyi*. Moscow: Kinofotoizdat, 1935.

Alpers, Boris. *Dnevnik kinokritika: 1928–1937*. Moscow: fond Novoe tysiacheletie, 1995.

Blok, Alexander. *Sobraniye sochinenii*, vol. 7, edited by Vladimir Orlov, Alexander Surkov, Kornei Chukovskii. Leningrad: Gosizdatel'stvo khudozhestvennoi literatury, 1963.

Bulgakowa, Oksana. *Golos kak kul'turnyi fenomen*. Moscow: Novoe Literaturnoe Obozrenie, 2015.

Eisenstein, Sergei. *Metod/Die Methode*, edited by Oksana Bulgakowa. Berlin: PotemkinPress, 2009.

Eisenstein, Sergei, Vesvolod Pudovkin, and Grigori Alexandrov. "Statement on Sound," Sergei Eisenstein. *Writing: 1922–34*, edited and translated by Richard Taylor, 113–14, London: I. B. Tauris, 2010.

35 Quoted in Knebel, *Vsia zhizn'*, 109–10.

Gardin, Vladimir. "Kak ia rabotal nad obrazom Babchenko." *Sovetskoe kino*, no. 1–2 (1934): 96–100.

Helmholtz, Hermann von. *Die Lehre von den Tonempfindungen als Physiologische Grundlage für die Theorie der Musik.* Hildesheim: G. Olms, 1968.

Janecek, Gerald. *Zaum: The Transrational Poetry of Russian Futurism.* San Diego: San Diego State University Press, 1996.

Kaganovsky, Lilya. *The Voice of Technology: Soviet Cinema's Transition to Sound, 1928–1935.* Bloomington: Indiana University Press, 2018.

Kalganov, Yuri. "Pervye gody Leningradskogo radio." In *Pervye gody sovetskogo muzykal'nogo stroitel'stva: Stat'i: Vospominaniia: Materialy*, 205–230. Leningrad: Sovetskii kompozitor, 1959.

Knebel, Maria. *Vsia zhizn'.* Moscow: VTO, 1967.

Kozintsev, Grigori. *Sobranie sochinenii*, edited by Iakov Butovskii and Valentina Kozintseva. Leningrad: Iskusstvo, 1982.

Montagu, Ivor. *With Eisenstein in Hollywood.* New York: International Publishers, 1969.

Olenina, Ana Hedberg. *Psychomotor Aesthetics: Movement and Affect in Modern Literature and Film.* New York: Oxford University Press, 2020.

Piotrovskii, Adrian. *Teatr: Kino: Zhizn'.* Leningrad: Iskusstvo, 1969.

Posener, Valérie. "To Overtake and Outstrip Hollywood: Early Talking Pictures in the Soviet Union." In *Sound, Speech, Music in Soviet and Post-Soviet Cinema*, edited by Masha Salazkina and Lilya Kaganovsky, 60–80. Bloomington: Indiana University Press, 2013.

Pudovkin, Vsevolod. *Sobraniye sochinenii*, edited by Tatiana Zapasnik and Ada Petrovich. Moscow: Iskusstvo, 1974.

Radek, Karl. "Dve fil'my." *Izvestiia*, April 23, 1931, 4.

RGALI (Russian State Archive of Literature and the Arts), Moscow.

Smirnov, Andrey. *Sound in Z: Experiments in Sound and Electronic Music in Early 20th Century Russia.* London: Koenig Books, 2013.

Taylor, Richard, and Ian Christie, eds. *The Film Factory: Russian and Soviet Cinema in Documents 1896–1939.* New York: Routledge, 1988.

Thompson, Kristin. "Early Sound Counterpoint." *Yale French Studies*, no. 60 (1980): 115–140.

Tsekhanovskii, Mikhail. "Dela i liudi kak professional'nyi urok." *Proletkino*, no. 19/20 (1932): 15–16.

Widdis, Emma. "Making Sense without Speech: The Use of Silence in Early Soviet Sound Film." In *Sound, Speech, Music in Soviet and Post-Soviet Cinema*, edited by Masha Salazkina and Lilya Kaganovsky, 100–116. Bloomington: Indiana University Press, 2013.

About the Author

Oksana Bulgakowa is retired Professor of Film Studies at Johannes Gutenberg University in Mayans. Among her book publications are *Sergej Eisenstein: A Biography* (2002) and *Resonanz-Räume: Die Stimme und die Medien* (2012). Recent articles include "Vocal Changes: Marlon Brando, Innokenty Smoktunovsky, and the Sound of the 1950s" in *Sound, Speech, Music in Soviet and Post-Soviet Cinema* (2018). She has also made films (e.g., *Die verschiedenen Gesichter des Sergej Eisenstein*, 1998, together with Dietmar Hochmuth), curated exhibitions (e.g., *Sergei Eisenstein: The Mexican* Drawings, 2009) and developed multimedia projects (e.g., *Factory of Gestures: On Body Language in Film* 2008; *Sergei Eisenstein: My Art in Life*, Google Arts & Culture, 2017).

III

National Contexts

7. Early Japanese Sound Film Aesthetics at Shochiku and Nikkatsu

Johan Nordström

Abstract: This chapter traces how the film studios Shochiku and Nikkatsu participated in the Japanese film industry's extended transition to sound cinema. Shochiku pursued a production policy that encouraged its directors to explore the potential of sound film aesthetics piecemeal, as they continuously went back and forth between silent cinema, sound-version films, and full talkies. Nikkatsu came to lead the way in redefining the aesthetics of the *jidaigeki* genre, gradually re-shaping the norms and aesthetic modes to bring contemporary aesthetics and thematic concerns into their representations of the past. This chapter explores how Shochiku and Nikkatsu negotiated issues such as songs, language, elocution and other sound practices, and the changing relation between on-screen and off-screen space.

Keywords: language, music, realism, period films, historical reception

For critic Ōtsuka Kiyoshi, the decisive factor behind the abundance of excellent Japanese films released in 1936 was the completion of the industry's transition to sound.[1] Indeed, when Ozu Yasujirō directed his first talkie *Hitori musuko* (*The Only Son*) (1936), he was the last of the major directors in Japan to do so, and it signalled the irrevocability of the Japanese transition. Although Japan's first technologically and critically successful talkie *Madamu to nyōbō* (*The Neighbour's Wife and Mine*, dir. Gosho Heinosuke) was released in 1931, sound film did not form the majority of films produced in Japan until 1935, and then only if we include the various iterations of the so-called *saundo-ban*, or "sound version" – silent films

1 Ōtsuka, *Nihon eiga*, 50.

Wiegand, D. (ed.), *Aesthetics of Early Sound Film: Media Change around 1930*.
Taylor & Francis Group, 2023
DOI 10.5117/9789463727372_CH07

with a mechanically synchronized soundtrack consisting of music, song, special effects, and sometimes unsynchronized dialogue/narration – which was a widespread practice during the Japanese cinema's transition from silent to sound.[2]

This protracted transition had primarily economical and structural causes. Financial constraints delayed investment in the necessary facilities and technology, and the Japanese film industry was geared toward the quick production of cheaply made products designed for rapid dissemination, in contrast to the more expensive, high-end product that sound films – said to cost three times more to produce than silents – constituted at the time.[3] Furthermore, the uneven spread and quality of sound recording and playback technology within the industry and along regional divides, ensured that the transition continued throughout the 1930s, with *benshi* narration accompanying many rural screenings of sound films well into the 1930s and silent films still being produced as late as 1941. However, during the latter half of the 1930s, transitional aesthetics and screening practises were no longer an industry-wide phenomenon, but instead emblematic of financially weaker, second-rung studios and peripheral, local, screening practices.

During the transition several new production companies that focused on sound film emerged. Some, like Orientaru eiga-sha, Shineigasha or Ongageijutsukenkyujo, became short-lived, whereas Tokyo-based P.C.L., the Kabushiki Kaisha Shashin Kagaku Kenkyūjo, and Kyoto-based J.O. Studio, perhaps best illustrate those that successfully carved out a niche for themselves in the early sound film market.[4] P.C.L., especially, geared its production initially towards musically infused up-beat light entertainment, drawing inspiration from early Hollywood talkies, musicals and German operetta films.[5] This chapter will examine the use of sound and picture in several representative works by Shochiku and Nikkatsu, and correlate this with contemporary theoretical and critical discourses. This will highlight similarities and discrepancies in the approaches of well-established production companies and filmmakers as they adapted in order to meet the demands of sound cinema.

2 For a detailed analysis and data on the production of sound-version films, see Nordström, "Between Silence," 217.

3 Kinoshita, "Benshi track," 6.

4 The studios would merge in 1937 to form the two main parts of the Tōhō Film Company.

5 For a detailed analysis on P.C.L.'s early sound film output, see Nordström, "Modern Talkie," 157–81.

Shochiku

While in New York, Shochiku's head of production, Kido Shirō, observed Warner's Vitaphone system, as well as the current state of RCA and Western Electric sound systems.[6] Upon his return to Japan in 1929, he quickly established a small 'sound film' research unit and hired two technicians, the Tsuchihashi brothers, who utilized Western technical journals to help them design Shochiku's so-called "pure Japanese technology,"[7] the eponymous Tsuchihashi Shochikuphone system, first utilized by Shochiku for *Madamu to nyōbō*, and later also adopted by Shinkō Kinema.

With Shochiku's transition to sound, 'Kamata modernism,' the distinctive vibrant style of Shochiku's films, gradually gave way to new stylistic modes of storytelling. Many of the studio's directors simplified their styles, abandoning avant-garde techniques with rapidly changing and unexpected camera angles, opting instead for realistic narratives and a more subtle cinematic technique. These changes helped to cement the tradition of understated realism which gradually became a central mode in Japanese film art during the 1930s. The necessity of direct sound recording also became a technological prerequisite for abandoning the use of loud arc lights in favour of incandescent tungsten white light bulbs, which, as detailed by Daisuke Miyao, came to have wide reaching influence on not only Shochiku's visual style, but across the Japanese film industry as a whole.[8] Shochiku's early sound films also exhibit a shift in narrative focus towards a form of 'everyday realism,' or stories that were predominantly in the vein of light-hearted melodrama and popular comedy, rooted in the sounds of everyday life. These early talkies came to focus on family life, the intimacy of voices, the act of listening, and misunderstandings, in contrast to the more sophisticated silent film literary adaptations, and socially aware *shoshimin-geki* melodrama, that directors such as Shimazu Yasujirō, Ushihara Kiyohiko, Shimizu Hiroshi, and Ozu Yasujirō otherwise were associated with in the 1930s.

Although it was preceded by several Japanese sound film productions, it is Gosho's *Madamu to nyōbō* that has become widely recognized as Japan's first 'authentic' (*honkakuteki*) talkie (Fig. 7.1). Narratives of artistic or technological 'firsts' always risk sacrificing historical complexity and nuance for simplicity, yet *Madamu to nyōbō* stands out as the first Japanese talkie

6 Kido, *eigaden*, 69, 84.

7 Tsuchihashi, "Eiga ga koe," 300. Unless noted otherwise, all translations are mine.

8 For an exhaustive exploration of the use of light and its influence on Shochikuis style, see Miyao, *Aesthetics of Shadow*.

to win both broad commercial success and unequivocal critical praise, ranking as number one in that year's *Kinema junpō* critics' poll. At heart a *nansensu-eiga*, or 'nonsensical comedy' – a genre that Gosho excelled at – *Madamu to nyōbō* was praised for its imaginative use of off-screen sound and its naturalistic style of acting and elocution. *Madamu to nyōbō* had the apt working title *Tonari no zatsuon* (*The Noise Next Door*), as it details the travails of a playwright (Watanabe Atsushi), who is further distracted by several (off-screen) noises, including his wife (Tanaka Kinuyo as an archetype of the traditional Japanese woman) and children in the adjacent room, mice scurrying in the attic, a tomcat outside his window, and more egregiously jazz music emanating from the house next door. As he goes next door to complain, he instead gets a free drink and becomes placated by the sex appeal of the *moga*, 'modern girl' (Date Satoko), in charge of the jazz ensemble. Revitalized by the up-tempo of the jazz song "Supīdo hoi" ("Speed Up!") that the band is rehearsing,[9] he eventually returns home to briskly finish his draft before the deadline.

Gosho makes frequent use of sound devices such as off-screen noise and dialogue as well as sound perspective for plot development and comic relief, sometimes revealing the source of the sound in ensuing scenes, at other times keeping the sound as a purely aural presence within the diegesis. Iwamoto Kenji has convincingly argued that *Madamu to nyōbō* is "the Japanese film that most closely approximated its Western counterparts in both its detailed montage and its use of 'invisible' sounds to represent perspective" and locates the film in "the style of 'everyday realism,' which is closer [than other early Japanese sound films] to the French film but unlike the musically structured films of René Clair."[10] As Tamura Yukihiko writes in his 1931 review of *Madamu to nyōbō*, "With the exception of Watanabe Atsushi, none of the cast had experience on the stage, which actually works in their favour, as lines are spoken in an extremely natural manner. Ensuring that dialogue is delivered the same way as everyday conversation should be the most important issue for producers of Japanese talkies."[11] In *Madamu to nyōbō* the everydayness of the dialogue and natural elocution with which it is delivered adds a layer of authenticity to the film, in which Tanaka Kinuyo's dialect stood out against the standard Tokyo-based vernacular, and

9 The film's other theme song is called "The Age of Speed." Both songs were composed by Takashina Tetsuo and Shimada Haruyo and released on phonograph record in connection to the film's release.

10 Iwamoto, "Japanese talkies," 326.

11 Tamura, "Madamu to nyōbō," 77.

Figure 7.1: *Madamu to nyōbō* (1931, dir. Gosho Heinosuke).

according to Satō Tadao, endowed her character with a certain "coquettish eroticism."[12] As Kerim Yasar has shown, when "talkies started being produced in Japan, the project of language unification and modernization was, in theory at least, well advanced,"[13] a fact which has considerable implications for our understanding of the contemporary discourses concerning dialogue and elocution that erupted in the wake of the talkies. Gosho himself commented many years later, "When we cast Tanaka Kinuyo, who speaks with a Kansai accent, we broke the rule stipulating that all dialogue had to be delivered in standard Japanese, the way radio announcers speak. Contrary to expectations, her delivery imparted lifelike nuances to the dialogue and offered a valuable suggestion for solving problems faced by the talkies that followed."[14] Implicit in the discourses regarding dialogue and elocution in the talkies lies what Yasar has delineated as "the tension between the desire to use new auditory media as vectors of language standardization, and the potential to use them as showcases of linguistic differences."[15] As we shall see, this was especially so in the case of the period film, for which Yasar has

12 Satō, *Nihoneigashi*, 330.
13 Yasar, *Electrified Voices*, 217.
14 Gosho, "Madamu to nyōbō," 48.
15 Yasar, *Electrified Voices*, 198.

convincingly argued that the introduction of sound and the necessity for historically inflected dialogue resulted in a "crisis of authenticity,"[16] which came to tie in to a larger discourse on realism and perceived historical accuracy within the period film.

At Shochiku, Kinugasa Teinosuke had successfully explored how to effectively utilize off-screen sounds and dialogue with the short *jidaigeki* sound film *Ikinokatta Shinsengumi* (*Survivors of the Shinsen Group*) (1932), the first talkie to be shot on Shochiku's newly constructed sound stage at their Shimokamo studio, yet it was his adaptation of *Chūshingura*, the famous fictionalized account of the loyal forty-seven *rōnin* and their mission to avenge the death of their master, that would come to be seen as epoch-making. Director Kinugasa stated at the time that he was "driven by an intense desire to turn *Chūshingura*, a tale more familiar to Japanese readers and audiences than any other, into a great artistic work" and that he "was passionately determined to bring the stagnant *jidaigeki* back to life by evolving an aesthetic of the *jidaigeki* sound film."[17] The resulting *Chūshingura* (1932), which was ranked second best film of the year by the film journal *Kinema junpō*,[18] is indeed rich in examples of how Kinugasa exerted himself to effectively utilize off-screen sounds and sound bridges in scene transitions.

Chūshingura, with its intricate plot and extensive character gallery, offered numerous starring roles, and as it drew on talents from all of Shochiku's production units, Kinugasa had to work with actors speaking in a wide variety of different local dialects. Director and screenwriter Kiyose Eijirō had argued the previous year that divergent regional and theatrical backgrounds, each with their own style of elocution, would require standardization.[19] Yet Kinugasa chose a half-way approach: he modernized the language and instructed actors to speak naturally, and then utilized different dialects for different types of scenes, steering those actors with particularly strong dialects away from serious dialogue-driven scenes. At the time, Kinugasa defensively stated "I did the best I could,"[20] yet the modern dialogue was criticized and Kinuyo Tanaka's Kansai dialect, the very one that had been seen to add authenticity and sensuality to *Madamu to nyōbō*, was here seen as out of place.[21] This indicates that different standards were applied to

16 Ibid., 223.
17 Misono, *Eiga Chūshingura*, 63.
18 Ibid., 62.
19 Kiyose, "Tokii jidai-geki," 100–101.
20 Misono, *Eiga Chūshingura*, 63–64.
21 Wadayama, "*Chūshingura*," 163.

the *jidaigeki* talkies, and we shall return to the question of dialogue and elocution within the period film when we take a look at the films produced at Nikkatsu.

Even after the success of *Madamu to nyōbō,* Kido was keenly aware that Shochiku was far away from being able to fully commit to sound film production, and that the transition would necessitate the displacement of the cinema musicians and *benshi.*[22] The cinema musicians went first, gradually replaced with accompaniment on record and eventually on the soundtrack, and when the highly publicised industry-wide *benshi* strikes broke out in 1932, Kidō responded by increasing Shochiku's production of sound-version films, a strategy that he later described as instrumental in gradually pushing out the *benshi* and any remaining musicians.[23] Although an industrywide phenomenon, perhaps no other studio utilized the sound-version films as skilfully as Shochiku. Constituting an integral part of the studio's transition to sound, the sound-version production at Shochiku functioned on multiple levels. Most significantly, for the studio, such films constituted a stopgap as the studio prepared to convert to full-talkie production, padding the studio's roster of 'sound film' releases, while for directors working at Shochiku, sound-version production instigated an increased awareness of how sound effects and music might be used on the soundtrack, paving the way for their artistic transition to full-talkie sound film.

Shochiku veteran Shimizu Hiroshi's first full talkie (and eighty-sixth film altogether) *Nakinureta haru no onna yo* (*The Lady Who Wept in Spring*) was released in 1933, and he came to move with ease between directing silent films, sound-version films, and full talkies, shooting his last silent film *Tokyo no eiyu* (*A Hero of Tokyo*) in 1935. For Shimizu, famous for his lengthy tracking shots and sophisticated geometrical compositions, *Nakinureta haru no onna yo* was in many ways something of a departure from his usual visual style, yet its melodramatic narrative set in Hokkaido about the love story between an itinerant woman and a miner bridges the thematic concerns present in his silent films and more mature sound films. Shot partly on location in Hokkaido, Shimizu contrasts the snowy landscapes with the interior of the local tavern with a bar downstairs and room for the women working in the tavern on the second floor. Although Shimizu has simplified his geometrical compositions and reined in his use of tracking shots, possibly a result of the physical limitations of the sound technology, he skilfully explores the interiors of the tavern, while engaging in stylistic experimentation with

22 Kido, *eigaden*, 72.
23 Ibid., for a detailed analysis of the *benshi* strikes see Raine, "No Interpreter," 127–56.

light and shadow. He utilizes songs on multiple occasions, as well as off-screen sounds and voices; at a climatic moment near the end of the film, the protagonist faces off against the bullying foreman outside in the snow, and Shimizu situates a large part of their brawl off-screen, behind a wall of snow that blocks them up to their shoulders. Instead of showing the fight directly, he utilizes off-screen voices and sounds to depict it aurally and build tension between the visual fragments, such as shots of falling snow and snow-covered pine tree branches.

Although Shochiku continued with its policy of creating sound-version films and talkie films in tandem, it is evident that directors moving between the two modes of storytelling were quickly developing a refined sensibility for utilizing sound in cinema. Gosho continued to direct sound films that explored the appeal of intimate voices and misunderstandings, and in *Hanayome no negoto* (*The Bride Talks in Her Sleep*) (1933) and *Hanamuku no negoto* (*The Groom Talks in His Sleep*) (1935) the plot revolves around the motif of spying, or listening in, on the unconscious sleep-talk of the bride and the groom respectively. These films exhibited a certain erotic appeal in their evocation of 'uncontrollable' voices. Especially in *Hanayome no negoto*, the promise of hearing the already erotically charged sleep-drunk voice of Tanaka Kinuyo was enough to turn the film into a success,[24] and critic Kitagawa Fuyuhiko stated that Tanaka's elocution was on a level different from the rest.[25] Similarly, the raison d'être of Nomura Hiromasa's *Ureshii koro* (*Happy Times*) (1933) seems to have been the joy of eavesdropping on a newlywed couple who, besotted with each other, keep exchanging words of intimate affection.

Perhaps due to these and similar films, Gosho argued in 1934 that the artistic maturation of the Japanese talkie and the development of a distinctive 'sound film sensibility' was held back partially because staff and directors working at most studios had to divide their attention between talkies and silent films.[26] He furthermore argued that because film studios dealt with the talkies' inherently high production costs by lowering their level of intellectual sophistication (thus appealing to the widest possible audience), Japanese talkies – at least those made by Shochiku – ended up being limited to the genre of melodrama.[27] Yet, it was this very genre of melodrama and 'everyday realism' that gradually

24 Shochiku, "Shōchiku shichijūnenshi," 274.

25 Kitagawa, "Hanayome no negoto," 94.

26 Gosho, "Nihon tōkī," 57–62.

27 Ibid.

came to develop into the sophisticated realist sound films of the second half of the 1930s.[28]

Nikkatsu

Although Nikkatsu was early on enthusiastic about the talkies, almost nothing survives of the studio's initial output. Their first foray into sound film production saw the company utilize the Japanese sound-on-disc (*disuku shiki tōkī*) technology Eastphone (*īsutofon*), for the period film *Hachisuka Koroku* (released in parts 1 and 2) (1929, dir. Takahashi Hiroyasu) and the contemporary drama *Ukina zange* (*Confession of a Scandal*) (1929, dir. Saegusa Genjirō). Even though both films contained audible dialogue, they were post-synchronized, and Nikkatsu seems to have been aware that this process left much to be desired. In the theatre programme for *Hachisuka Koroku*, the company notes that in comparison to foreign talkies this was an "inferior product," while asking for it to be seen from the point of view of a first attempt as well as a film that was "more than just an imitation of Western talkies."[29] Indeed, these films were scathingly criticized as technically and aesthetically inferior products, severely lacking in terms of sound quality. *Hachisuka Koroku* was described as "grotesque"[30] and its sound labelled as "noisy"[31] and "metallic,"[32] whereas *Ukina zange*, although seen as an improvement over *Hachisuka Koroku*, was criticized for its silent film aesthetics, especially its utilization of intertitles, which were seen to get in the way of the dialogue.[33]

After these arguably failed first attempts, Nikkatsu initiated a collaboration with the smaller talkie film production company Hassei Eiga Kabushiki Kaisha, marketed under the name of "Mina Talkie," which utilized Lee de Forest's Phonofilm sound-on-film technology. The collaboration culminated in the co-production of *Fujiwara Yoshie no furusato* (*Fujiwara Yoshie's Hometown*, hereafter *Furusato*) (1930, dir. Mizoguchi Kenji), featuring Fujiwara Yoshie, Japan's most popular singer at the time, in the lead role together with Nikkatsu's top ranked actress Natsukawa Shizue playing his romantic interest. Originally envisioned as a full talkie, the film eventually emerged

28 Iwamoto, "Japanese Talkies," 326.
29 "Kanda Nikkatsu puroguramu."
30 Sakae, "Shineigahyo," 6.
31 Tomoda, "Hachisuka Koroku," 100.
32 Ōkoku, "Hachisuka Koroku," 20–21.
33 Toshio, "toki kōgyo," 72.

as a part-talkie due to the technical limitations of Mina Talkie's Phonofilm technology and studio facilities that hadn't been soundproofed, as a result of which shooting had to be stopped every time a large truck or street vendor passed by outside. Although *Furusato* achieved commercial and critical success, most of its production staff were apologetic; for instance Mori Iwao, who co-wrote the original script, saw its hybrid nature as regrettable, and argued that ideally there should be a break between silent and talkie aesthetics.[34] Ironically, it was critically lauded for the way that its refined silent-cinema aesthetics imbued the film with a sense of fluidity, often otherwise lacking in early talkies.[35] In recent scholarship, Chika Kinoshita has highlighted the film's frequent and expressive use of parallel montage within its narrative structure,[36] whereas Nagato Yohei has argued that the film includes a richly organic use of off-screen sounds, and that its subtle sound design utilizes a concept of "counterpoint" and a technique of "cutaway within the shot" that anticipates the long take style that came to dominate Mizoguchi's later work.[37]

After Shochiku's critical and commercial success with *Madamu to nyōbō*, Nikkatsu was spurred to return to sound film production in 1932, initiating a collaboration with the newly formed sound studio P.C.L., the Kabushiki Kaisha Shashin Kagaku Kenkyūjo ("Photo Chemical Laboratory Stock Company"). On Nikkatsu's behest P.C.L. utilized their newly developed sound-on-film recording system P.C.L. shiki ("P.C.L. System"), to add so-called *afureko* (sound that was recorded separately or later) as well as direct sound recording to several films marketed as all-talkies, starting with the simultaneous release in April of *Onna Kunisada* (*The Female Kunisada*) (1932, dir. Kiyose Eijirō), and *Toki no ujigami* (*Timely Mediator*) (1932, dir. Mizoguchi Kenji), both released in talkie and silent versions.

Toki no ujigami was based on a popular novel by Kikuchi Kan. Mizoguchi stated that "I wanted to make this film into a Japanese operetta, and if we had had the time and money I would have liked to add a theme song and good musical accompaniment as well. Sadly that proved impossible, so instead I tried to add as many singing people and as much [diegetic] music to the street scenes as possible."[38] Yet it was this use of music and song that garnered the film's most scathing criticism; it was seen as severely

34 Mori, "Furusato zadankai," 53.
35 Iijima, "Furusato," 89.
36 Kinoshita, "Mizoguchi," 177–85.
37 Nagato, "Hometown," 183–99.
38 Sasō, "5. Toki no ujigami," 93.

misjudged, and labelled "sleazy," "vulgar," and "base" to the point that it became insulting to the audience.[39] The quality of the sound was generally praised, although still seen to be too "mechanical" and "lacking in human warmth." One reviewer concluded that "even though you can feel the 'sound' in the film, as a talkie, it failed to stop you from feeling silence."[40]

Nikkatsu continued their collaboration with P.C.L. throughout 1932 and released several more sound films. These films varied in content, from melodramas such as *Haru to musume* (*Spring and a Girl*) (1932, dir. Tasaka Tomotaka), to nonsensical comedies, and period films such as *Tabi wa aozora* (*Travels Under the Blue Sky*) (1932, dir. Inagaki Hiroshi), the first to receive partial simultaneous sound recording by P.C.L. The *Kinema junpō* review praised *Haru to musume* for its talkie aesthetics and the way the soundscape matched the film's mood.[41] Similarly, in an article discussing the use of sound effects in contemporary talkies, *Haru to musume* was praised for its creative use of sound effects and sophisticated sound bridges, such as when a young woman struggles to get onto a train and her heavy breathing merges with that of the noise from the steam train.[42] By contrast, Inagaki was criticized for being stuck in the aesthetics of the silent film, as the use of sound and sound effects in *Tabi wa aozora* only reached the most basic level, such as the sound effect for a nail being hit by a hammer, while resorting to camera work and visual effects in order to illustrate psychological states.[43] Furthermore, the quality of the sound from the P.C.L. sound system was once again criticized. Contemporary critic Ishigami Iku concludes that although it might be fine for one of the first talkies, the lack of sound quality keeps the film from being anything more, "so that it would have been better to have made it into a silent film."[44]

Tabi wa aozora is notable for being the first Japanese period film to utilize modern colloquial language and elocution on its soundtrack. Inagaki, one of the leading reformers of the *jidaigeki* in the 1930s, was inspired by American and European cinema as well as the contemporary Japanese *shomingeki* films, and famously argued that *jidaigeki* should be seen as *chonmage wo tsuketa gendai-geki,* or "contemporary dramas with topknots"; the idea being that period films should utilize modern forms of expression, so as to be indistinguishable from contemporary films about common people except

39 Ibid.
40 Wadayama, "Toki no ujigami," 39.
41 Wadayama, "Haru to musume," 39.
42 Shigeno, "Onkyō," 68.
43 Ibid.
44 Ishigami, "Funman nihon tōkī," 39.

for setting and costumes.[45] However, Inagaki's use of modern colloquial language and elocution came to galvanize the critical community. Shimizu Masao's writing is representative of those that were critical: he argued that part of the essence of the period film lies in how faithfully it adheres to historical reality, and he saw the introduction of modern dialogue as directly opposed to this. Arguing that the movie audience would surely rebel against such blatant transgression, he singles out *Tabi wa aozora* as an experiment that embarked upon the wrong path for the future period film, stating that "it would be better to call it a *gendaigeki* [...], since it has lost the essence of the period film."[46] Others, like Kitagawa Fuyuhiko, while remaining critical of *Tabi wa aozora*, nevertheless argued for the necessity of creatively utilizing modern speech in the period film.[47]

The early 1930s were a period of economical and structural turmoil for most of the film studios, and for Nikkatsu these troubles reached a nadir in a series of strikes and lay-offs in 1932. Much of the turmoil was connected to the coming of sound cinema, for instance the *benshi*'s and musicians' anti-talkie strikes and Nikkatsu's massive layoffs and attempts to cut staff salaries in order to afford to equip their cinemas for sound film.[48] Although Nikkatsu had signed a contract with P.C.L., who on the behest of Nikkatsu had begun the construction of their new, modern sound studio facilities, Nikkatsu reneged on the contract. Instead, they built their own sound film studio at Tamagawa and started to produce sound films in-house in the hope of turning around their fiscal deficit, while simultaneously investing in the Western Electric sound technology as it offered both recording and playback technology.[49] Western Electric was seen by the Japanese film industry as the gold standard of sound systems, and Nikkatsu came to describe their acquisition of it as a great triumph, later stating that "without a doubt, Western [Electric], the proud world champion [of sound systems], became the Nikkatsu talkie's greatest strength."[50] 1933 saw the return of

45 Yamamoto, "Inagaki Hiroshi," 31.

46 Shimizu, "Jidaieiga shōron," 45–47.

47 Kitagawa, "Jidaieiga," 63.

48 The Great Nikkatsu Strike occurred in August 1932 and involved more or less the whole studio. It resulted in 180 production staff lay-offs, and studio head Ikenaga Kōkyū retiring from his post. In its aftermath, the so-called Shichiningumi incident occurred on 20 September, when seven of Nikkatsu's most famous remaining staff and directors decided to leave the studio to form the Shineigasha. The fallout from the turmoil saw Nikkatsu run a deficit and their film production severely curtailed.

49 See Ōmura, "shikikisho (2)," 7–9, for the deficit plans and Mori, "*Watakushi*," 157, for the investment.

50 Nikkatsu, *Yonjunenshi*, 100.

many actors and directors who had left Nikkatsu during the strikes, and although the studio output was predominantly silent, Nikkatsu did release six talkies; four period films and two modern dramas. This increased to twenty-three in 1934, and finally in 1935 Nikkatsu ceased production of silent films completely, even though they still produced a few sound-version films.

Of the six talkies Nikkatsu released in 1933, it was the period film *Tange Sazen dai-ippen* (*Tange Sazen, First Part*), directed by sword film virtuoso Itō Daisuke, famous for his fluid camera, fast cutting style and directorial speed, tempo, and rhythm, that received the most attention. Even though *Tange Sazen* (named after the famous fictional swordsman) was shot as an *afureko*, as Nikkatsu still lacked soundstages, its sound quality was praised and the critical consensus seems to have been one of deep appreciation. Itō was lauded by Nakata Kameinosuke for not merely creating a "silent film plus sound," but instead striving for new artistic methods for the merging of sound and picture.[51] Nakata compared Itō with Kinugasa, who was criticized for merely being proficient in the technological aspects of adding sound to film and having yet to develop a proper sound film style.[52] Iida Shinbi in his review of the film praised the way Itō had succeeded in bringing his characteristic strengthes from the silent film to the talkie, giving examples of how Itō would skilfully use sound bridges, like the sound of a gardener's scissors cutting to the use of scissors inside a house, or the soft drumming sound of a Japanese-style back massage being carried over to the sound of clapping at a banquet. The repeated and varied use of theme songs, such as a character starting to sing a song that turns halfway through into the whistling tune of another character in the next scene, was also noted.[53] Like Iida, Kitagawa Fuyuhiko argued that Itō had succeeded in bringing his silent aesthetics to sound, by adapting his scene changes from the matching of visual shapes to that of sounds.[54] Indeed, some saw *Tange Sazen* as a "kind of renaissance, signalling a new phase for Itō Daisuke."[55]

Critic Ōtsuka Kiyoshi was generally positive towards the film, praising Itō for the way he gathered strong performances from his cast, most of whom were still not used to delivering dialogue, but also lamented how Itō's sophisticated visual style, so iconic of his silent films, wasn't as thoroughly developed in *Tange Sazen*.[56] The issues concerning *jidaigeki* dialogue and

51 Nakata, "geijutsuron," 176.
52 Ibid.
53 Iida, "Tange Sazen," 48.
54 Kitagawa, "Toritomenonai kanso," 72.
55 Tomonori, "Itō Daisuke," 120.
56 Ōtsuka, *Nihon eiga*, 84–89.

elocution still remained. During a round-table discussion, Itō spoke of the difficulties in achieving uniformity of elocution, citing the chaotic delivery of lines resulting from the actors' divergent backgrounds within various theatrical traditions, oral storytelling traditions, and film, as well as unclear pronunciation due to regional accents.[57] "I would really like to see the creation of a new manner of period film elocution. I intend to work on it, too, but I think that it is our most pressing need."[58] Yet, Itō argues that dialects can be utilized in such a way as to add depth to characters, for instance, giving Osaka-dialect to a character he felt thin, or removing colourful dialects from characters he wanted to beautify.[59]

Eventually, the scriptwriting collective Narutaki-gumi,[60] in their effort to modernize the period film, would tackle the issues of dialogue head on. Inagaki would recall how "the biggest task facing the group was the modernization of the language, stating that there were writers who couldn't write anything besides the *benshi*'s seven-five syllable metre, and actors who couldn't utter modern language. Thinking back now to how Narutaki-gumi's work attacked the obstinate narrow-minded critics set against modern language, I believe we succeeded due to our youth and camaraderie."[61]

Eventually, it would be Inagaki who, together with fellow Narutaki-gumi member, screenwriter Mimura Shintaro, came to formalize what would become the standard form of delivery for period film dialogue: samurai were to speak like white-collar office workers in the upmarket Marunouchi district of Tokyo, and travellers, gamblers, and gangsters were to use the blue-collar idiom of the capital's poorer Shitamachi area.[62]

Conclusion

In response to the new possibilities of sound cinema, the Japanese film industry's extended transition facilitated aesthetic experimentation and a

57 Itō, "Jidaigeki tōkī zadankai," 203.

58 Ibid., 207.

59 Ibid., 203.

60 Active between 1934 and 1938, Narutaki-gumi consisted of eight screenwriters and directors affiliated with different studios: Inagaki Hiroshi, Yamanaka Sadao, Takizawa Eisuke, Fujii Shigeji, Hagiwara Ryō, Yahiro Fuji, Mimura Shintarō, and Suzuki Momosaku. For a detailed analysis of the group and its impact, see: Iris Haukamp, "The Group of the Roaring Waterfall: Researching the Narutaki Gumi Filmmaking Collaborations (1934–1937)," *Journal for Japanese Studies* 10 (2020): 153–70.

61 Inagaki, "*Nihon*," 124.

62 Yamamoto, "*Inagaki*," 31.

gradual shifting of representational styles and thematic concerns. Shochiku, under the guidance of Kidō Shiro, came to pursue a production policy that for practical reasons came to encourage its directors to explore the potential of sound film aesthetics piecemeal, as they continuously went back and forth between silent cinema, sound-version films and full talkies. As Shochiku's early talkie style of 'everyday realism' gradually evolved into the mainstream realistic sound films that were to characterize the studios's output during the 1930s, within the period film, the transition to sound instigated a crisis of authenticity for the genre as a whole, centred on the question of dialogue and elocution.

Nikkatsu led the way in redefining the aesthetics of the *jidaigeki* genre, gradually re-shaping its norms and aesthetic modes to bring contemporary aesthetic and thematic concerns into their representations of the past. Although Nikkatsu was to continue to suffer economically throughout the 1930s, after having phased out its production of sound-version films in 1936, it would, under the guidance of studio head Negishi Kanichi and assistant manager Makino Mitsuo, come to enter a golden era of realist modern dramas, literary adaptations, war time pictures, and period films.

By the latter half of the 1930s, the major studios had solidified their sound film shooting and screening practises. Thus, the aesthetic issues negotiated by the early Japanese talkies, such as the incorporation of popular song, questions regarding language, elocution and other sound practises, and the changing relation between on-screen and off-screen space, gave way to a growing industry-wide interest in filmic realism that came to influence the Japanese cinema's aesthetics, such as the use of nondiegetic music in narrative film,[63] as well as its content.

Bibliography

Gosho, Heinosuke. "Madamu to nyōbō." In *Nihon eiga shinario koten zenshu vol.* 2, 48, Tokyo: Kinema Junpōsha, 1966.

Gosho, Heinosuke. "Nihon tōkī no danpenteki jōhō." *Ongaku sekai* 6, no. 6, (1934): 57–62.

Iida, Shinbi. "Nikkatsu jidaigeki tōkī Tange Sazen." *Kinema junpō*, November 12, 1933.

Ijima, Tadashi. "Furusato." *Kinema junpō*, April 21, 1930.

Inagaki, Hiroshi. *Nihon eiga no wakaki hibi*. Tokyo: Mainichi shinbunsha, 1978.

63 For a detailed analysis of the impact on nondiegetic music in narrative film, see Shibata, "Realism."

Ishigami, Iku. "Funman nihon take." *Kinema junpō*, August 11, 1932.

Itō, Daisuke, Īda Shinbi, Hazumi Tsuneo, Kobayashi Isamu, Kitagawa Fuyuhiko, Kisaragi Bin, Kishi Matsuo, Sugiyama Shizuo. "Itō Daisuke o kakonde: Jidaigeki tōkī zadankai." *Kinema junpō*, January 1, 1934.

Iwamoto, Kenji. "Sound in the Early Japanese Talkies." In *Reframing Japanese Cinema: Authorship, Genre, History*, edited by Arthur Nolletti, Jr. and David Desser, 312–27. Bloomington: Indiana University Press, 1992.

Kido, Shiro. *Nihon eigaden: Eiga seisakusha no kiroku*. Tokyo: Bungei shunjū shinsha, 1956.

Kinoshita, Chika. "The Benshi Track: Mizoguchi Kenji's the Downfall of Osen and the Sound Transition." *Cinema Journal* 50, no. 3 (2011): 1–25.

Kinoshita, Chika. *Mise-En-Scène of Desire: The Films of Mizoguchi Kenji*. PhD diss., University of Chicago, 2007.

Kitagawa, Fuyuhiko. "Hanayome no negoto." *Kinema junpō*, February 1, 1933.

Kitagawa, Fuyuhiko. "Jidaieiga ni okeru iwayuru 'wajutsu' nazo." *Kinema junpō*, August 1, 1933.

Kitagawa, Fuyuhiko. "Toritomenonai kanso." *Kinema junpō*, January 12, 1934.

Kiyose, Eijirō. "Tokii jidai-geki no daiyarogu." In *Eiga kagaku kenkyu*, no. 9 (September 1931): 100–101.

Misono, Kyōhei. *Eiga Torai nanajunen: Eiga Chūshingura*. Tokyo: n.p., 1966.

Miyao, Daisuke. *The Aesthetics of Shadow: Lighting and Japanese Cinema*. Durham: Duke University Press, 2013.

Mori, Iwao. *Watashi no geikai henreki*. Tokyo: Seigabo, 1975.

Mori, Iwao, Tanaka Eizō, Hatamoto Shūichi, Kisaragi Bin, Kitamura Komatsu, Mizoguchi Kenji, Mineo Yoshio, Tsutami Jōbu, and Ōno Motomu. "Furusato zadankai." *Eiga ōrai* 6, no. 63 (May 1930): 41–56.

Nagato Yohei. "The Dawn of the Talkies in Japan: Mizoguchi Kenji's Hometown." In *The Culture of the Sound Image in Prewar Japan*, edited by Michael Raine and Johan Nordström, 183–99. Amsterdam: Amsterdam University Press, 2020.

Nakata, Kameinosuke. "Hitotsu Toki geijutsuron." *Kinema junpō*, January 1, 1934.

Nikkatsu kaubushikikaisha. "Kanda Nikkatsu puroguramu." Accessed June 12, 2021, https://www.nikkatsu.com/movie/12752.html.

Nikkatsu kabushikikaisha. *Nikkatsu yonjunenshi*. Tokyo: Nikkatsu, 1952.

Nordström, Johan. "Between Silence and Sound: The Liminal Space of the Japanese 'Sound Version.'" In *Routledge Handbook of Japanese Cinema*, edited by Joanne Bernardi and Shota T. Ogawa, 213–30. London: Routledge, 2021.

Nordström, Johan. "The Image of the Modern Talkie Film Studio: Aesthetics and Technology at P.C.L." In *The Culture of the Sound Image in Prewar Japan*, edited by Michael Raine and Johan Nordström, 157–81. Amsterdam: Amsterdam University Press, 2020.

Ōkoku, Fujimoto. "Fugu no Nihon tōkī: Hachisuka Koroku o miru." *Eiga jidai* 6, no. 10 (October, 1929): 20–1.

Ōmura, Einosuke. "Ōmura Einosukeshi kikisho (2)." *Kiroku eiga,* no. 279 (February 1986), 7–9.

Ōtsuka, Kyoichi. *Nihon eiga kantokuron*. Tokyo: Eiga hyōronsha, 1937.

Raine, Michael. "No Interpreter, Full Volume: The *Benshi* and the Sound Transition in 1930s Japan." In *The Culture of the Sound Image in Prewar Japan*, edited by Michael Raine and Johan Nordström, 127–56. Amsterdam: Amsterdam University Press.

Sakae. "Shineigahyo 'Taii no musume.'" *Asahi Shimbun*, October 31, 1929.

Sasō, Tsutomu, and Nobuyoshi Nishida. "5. Toki no ujigaki." In *Mizoguchi Kenji: Jōen no hate no onna tachi yo genmu e no riarizumu*, edited by Tsutomu Sasō and Nobuyoshi Nishida, 93. Tokyo: Firumuatosha, 1997.

Satō, Tadao. *Nihon eigashi, vol. 1*. Tokyo: Iwanami shoten, 1995.

Shibata, Kotaro. "'Realism' in Late 1930s Japanese Films and Shiro Fukai's Accompaniment Music: Contemporary Arguments for the Rejection of Non-diegetic Music." *Aesthetics*, no. 21 (2018): 149–61.

Shigeno, Tatsuhiko. "Onkyō ni kansuru kansō." *Kinema junpō*, September 1, 1932.

Shimizu, Masao. "Jidaieiga shōron." *Kinema junpō*, September 11, 1932.

Shochiku kabushikikaisha. "Shōchiku shichijūnenshi." Tokyo: Shochiku, 1964.

Tamura, Yukihiko. "Madamu to nyōbō." *Kinema junpō*, August 21, 1931.

Tanaka, Masasumi, ed. *Shimizu Hiroshi: Sokkyōsuru poejī yomigaeru chō eiga densetsu*. Tokyo: Firumuatosha, 2000.

Tomita, Mika, ed., *Chie puro jidai: Kataoka Chiezō Inagaki Hiroshi Itami Mansaku shadatsu ni entāteinmento*. Tokyo: Firumuatosha, 1997.

Tomoda, Junichirō. "Hachisuka Koroku daininen." *Kinema junpō*, September 11, 1929.

Tomonori, Saiki, ed. *Itō Daisuke: Hangyaku no passhon jidaigeki no modanizumu*. Tokyo: Firumuatosha, 1996.

Toshio, Iwase. "Nihon ni okeru toki kōgyo no mirai." *Eiga Ōrai* 6, no. 60 (February 1930): 72–75.

Tsuchihashi, Takeo. "Eiga ga koe wo hatatsushita toki." In *Kikigaki kinema no seishun*, edited by Kenji Iwamoto and Tomonori Saiki, 285–316. Tokyo: Riburopōto, 1988.

Yamamoto, Kikuo. "Rensai ronbun: hikaku eigashi kenkyu (26): Amerika eiga to jiyushugi jidai-geki, Inagaki Hiroshi kantoku sakuhin o megutte." *FC Firum senta*, no. 56 (1979): 30–36.

Yasar, Kerim. *Electrified Voices: How the Telephone, Phonograph, and Radio Shaped Modern Japan, 1868–1945*. New York: Columbia University Press, 2018.

Wadayama, Shigeru. "Nikkatsu gendai eiga Haru to musume." *Kinema junpō*, June 21, 1932.

Wadayama Shigeru. "Toki no ujigami." *Kinema junpō*, April 21, 1932.

Wadayama, Shigeru. "Chūshingura." *Kinema junpō*, January 1, 1933.

About the Author

Johan Nordström is Associate Professor at the Department of Global Education, Tsuru University. He is co-editor of *The Culture of the Sound Image in Prewar Japan* (2020) and has written on various aspects of Japanese cinema's transition to sound. Recent articles include "Between Silence and Sound: The Liminal Space of the Japanese 'Sound Version'" (2020). Together with Alexander Jacoby, he has also curated several film programmes on Japanese silent and early sound cinema for Le Giornate del Cinema Muto, Pordenone and Il Cinema Ritrovato, Bologna. He is currently working on a book on the Tokyo based early sound film studio P.C.L., later Toho.

8. Early Sound Films in France: Contexts and Experiments

Martin Barnier

Abstract: Through an analysis of some of the first sound films produced in Paris, this chapter shows that even though contemporary critics frequently dismissed French films from this period as a return to 'canned theatre,' many films made by the major production companies were in fact highly experimental. Elaborate camera movements, inventive use of direct and off-screen sound, as well as complex editing patterns characterize these productions, especially at their beginnings. Moreover, some films self-consciously display sound film as a new medium, partly by addressing its relation to other sound media of the time. Only in 1934 experimentation seems to slow down when a more 'classical' way of using sound becomes common.

Keywords: film and theatre, visual style, camerawork, film editing, historical reception

At the beginning of the conversion to sound in France, several critics and filmmakers were furious about the talkies, attacking them as being static and dialogue-based. *This is canned theatre!* was a common complaint.[1] Jean Mitry remembered this moment later: "Any flexibility was lost. Many people were saying: 'The cinema is finished.'"[2] A journalist's reaction in a trade magazine from the summer of 1928 is exemplary of this kind of critique:

> The talking film is a contradiction in terms full of dreadful dangers. To add speech to silent images is [...] to completely distort the meaning of the screen. After thirty years of development, cinematographic technique

1 See, for instance, Aveline, "Présentation – Muet sonore ou parlant?" 17–20.

2 Mitry, "Entretien," 18–23 (unless otherwise noted, all translations are my own).

Wiegand, D. (ed.), *Aesthetics of Early Sound Film: Media Change around 1930.*
Taylor & Francis Group, 2023
DOI 10.5117/9789463727372_CH08

has acquired, today, a perfect eloquence that leaves nothing to be desired and which must borrow nothing from the eloquence of the word. [...] It is the actors' performance that loses its power, since the voice will attenuate a gesture and depict a state of mind; a whole range of techniques, from the most classical to the most audacious, vanishes in the face of the brutal sentence that used to be suggested in a tangible way; it is all the merits of the cinema, all its transcendent originality, that are overturned, denatured, disfigured, repudiated.[3]

The author of this article was one of many critics who had only seen talking pictures in London. In the summer of 1928, not one feature film with synchronized dialogue had been screened in France. Based on these experiences abroad, many French cinephiles feared the end of the acrobatic cinematography and extravagant montage preferred by the French avant-garde of the period. Filmmakers were also afraid of the "imperialism of the English language around the world,"[4] as director Marcel L'Herbier put it – a few months later, the famous avant-garde filmmaker agreed to direct talkies in French.

Another aspect that worried many critics was the idea of having noises reproduced faithfully in films. As Jacques Henri-Robert put it:

> Charm is always broken by noise, that annoying plague of our modern existence. With sound films, we will lose one of the last refuges where we could experience artistic emotions in a relaxing calm. For we must not hide the fact that the danger is inevitable: too much money is being conjured up to make noisy films for us to hope to avoid them. Whether we like it or not, in the near future, all films will be sonorous.[5]

Henri-Robert was right: audiences did not react as negatively as the critics. On the contrary, by the end of 1929, the talkies were already enormously successful in France. At first, these were American films like *The Jazz Singer* (dir. Alan Crosland, released in Paris in February 1929) or *Broadway Melody* (dir. Harry Beaumont, released in Paris in November 1929) that French audiences really appreciated. In October 1929, sixteen movie theatres were wired for sound in Paris, and a few more in other big French cities. By the end of 1930, there were 312 movie theatres wired for sound in France, including seventy

3 Vial, "Le danger du cinéma parlant."
4 L'Herbier, "Opinioni di artisti francesi sui films parlanti."
5 Henri-Robert, "Le film sonore."

in Paris. At the end of 1931, the number of cinemas equipped with sound technology in France had risen to 1,600 (out of a total of 4,000 cinemas).[6] In 1929, the first two feature films in French were made in London and released in France at the end of the year: *Les trois masques* (*The Three Masks*) (dir. André Hugon) and *La route est belle* (*The Road Is Fine*) (dir. Robert Florey).

Earlier short films with sound, such as the Gaumont Phonoscènes (of which 775 were produced between 1905 and 1915), had been largely forgotten by the late 1920s. With new technologies coming from Germany (Tobis/Klangfilm) and the USA (Vitaphone, RCA Photophone, Fox Movietone, etc.), French studios began mass-producing sound films in early 1930, which also affected the total number of feature films produced. Throughout the 1930s, over 110 films were produced each year in France (or with French producers), compared with fewer than sixty in the mid-1920s. Between 1929 and 1930, the French production of feature films increased by 207 per cent.[7] At the same time, the number of American films exported to France dropped. New independent producers such as Braunberger-Richebé, Marcel Vandal and Charles Delac, or Adolphe Osso took advantage of the new trend for talkies. Bernard Natan took over Pathé and quickly began making talkies too, as did the new Gaumont-Franco-Film-Aubert consortium. The majority of productions were mainstream films, referred to as "Saturday night cinema" (*cinéma du samedi soir*) at the time.

Were those films an artistic disaster like the critics were claiming? In what follows, I will take a closer look at some of the films' opening sequences, arguing that many popular French films from the conversion years in fact experimented heavily with the new sound technologies. In my analyses, I will focus on aspects such as camera movements, inventive sound/image juxtapositions (such as the use of off-screen sound), the use of direct sound, and inventive editing. Furthermore, I will highlight the self-reflexive display of the new medium of sound film and other sound media, as well as the films' self-conscious dissociation from 'canned theatre.'

Chiqué, Under the Roofs of Paris, and Other French Talkies from 1930

The first all-talking film produced and filmed in a French studio seems to be *Chiqué* (*Cheated*) by Pierre (sometimes spelled Pière) Colombier. Produced by Pathé-Natan with the American RCA Photophone system,

6 Billard, *L'Âge classique du cinéma français*, 29.

7 Barnier, *En route vers le parlant*, 108.

the (roughly) thirty-minute film was released on February 14, 1930. As one observer noted, "hundreds of spectators flocked to the Royal on Avenue de Wagram to see the first authentic 'talkie' from our country."[8] The film begins with an instrumental version of "J'embrasse votre main madame" by Ralph Erwin, who had acquired an international reputation in 1928 with the song's German version, "Ich küsse Ihre Hand, Madame" (with lyrics by Fritz Rotter), in the film of the same title starring Marlene Dietrich. In no time, one million records were sold.[9]

In *Chiqué*, the first camera movement occurs within the first shot. Moving backward from the orchestra, it encircles the dancers, giving the impression of being in the midst of the dancers. Thus, contrary to what critics feared, the camera is not static at all; instead, it gives the audience the feeling that they are participating in the party.

A similar beginning occurs in *Sous les toits de Paris* (*Under the Roofs of Paris*) (dir. René Clair) released by the French branch of the German Tobis Film company on April 28, 1930. René Clair later explained that he wanted to begin the film with a "tracking shot with sound" (*travelling sonore*) in order to make the audience feel as if they were gradually approaching the sound source.[10] Indeed, the film begins with an expansive camera movement towards a group of singers in a street of Paris (entirely built on a studio set) and a steady increase in volume on the soundtrack, conveying the idea of a sound perspective by means of a complicated camera movement. The rather fast camera movement was very much appreciated by critics: "Nothing drags. From the beginning, the film establishes the setting. A chimney, some smoke, houses from which a song seems to emerge: the houses grow, move forward, and, at the same time, in a parallel progression, the sounds of voices swell, become clearer: in a small square, a street singer has the crowd take up the chorus of the song he sells."[11]

8 "Chiqué."

9 Bennet, *Pianist Kartun and Composer Erwin*. Ralph Erwin composed songs for many French films, including "J'ai ma combine" for *Le Roi des resquilleurs* (*The King of the Gate Crashers*) (1930, dir. Pierre Colombier) and other successes. He lived in Paris after 1933. Over the years, there would be several more versions of this hit song in both German and French. In France, André Baugé, singer of the Opéra-Comique, was one of the most famous singers of the song, which was also recorded by Fred Gouin, Paul Gesky, Jean Sablon, and many others. In Germany, the Comedian Harmonists (among others), recorded the song in 1928 under the Electrola label, the German branch of His Master's Voice. On the significance of the radio in France during the 1930s, see Bennet, *La musique à la radio dans les années trente*, 314.

10 Marie, "*Sous les toits de Paris*, film de René Clair."

11 Girard, "Sous les toits de Paris."

The title song, which is repeated in the film, became a popular hit thanks to the film's audience, the buyers of sheet music, and radio listeners, emphasizing how much the conversion to sound was in fact a moment of media synergy: radio, records, and sheet music helped to publicize the film, which, in turn, helped to sell the song.

On August 1, 1930, Sofar-Film released *Prix de Beauté* (*Beauty Prize*), directed by Italian filmmaker Augusto Genina and starring American actress Louise Brooks. Originally a silent film shot in Joinville-le-Pont during the summer of 1929, it was later post-synchronized in several languages at the Tobis studio. Many scenes were shot on location. The film begins with documentary footage of bathers in a pool on Sunday. A tracking shot along bikes leads to the pool; an overhead panning shot shows the bathers having fun. We hear shouts and laughs as we discover the crowded swimming pool near the Marne River. Far away from a studio set, cinematographer Rudolph Maté films 'real life.' As one critic noted: "It is on the lively banks of the Marne, close to the Pont de Joinville, that some scenes of *Prix de beauté* were shot last week. The colourful bathing suits decorate the flesh browned by the sun in a funny way."[12]

Marcel L'Herbier's *Le mystère de la chambre jaune* (*The Mystery of the Yellow Room*) was released on September 8, 1930 by the producer and distributor Adolphe Osso. In contrast to L'Herbier's works from the 1920s, this is not an avant-garde film but a production quite in line with Osso's other mainstream productions. During the credits, suspenseful music accompanies the expressionist images. The high-contrast lighting by cinematographers Léonce-Henri Burel and Nikolai Toporkoff exaggerates the actors' frightened expressions. A dramatic voice-over presents the credits (the only written credits are the film title and the names of the producer and the director), making this one of the first French films experimenting with sound to introduce a new form of opening credits.[13] Significantly, the first technician to be named is Antoine Archimbaud, the sound engineer, depicted here with his RCA Photophone apparatus (Fig. 8.1). This is one of the first times – and maybe the last time – that a sound engineer is high-lighted

12 Mirbel, "Prix de beauté sur la plage de Joinville."

13 Another such film was the very succesful *Chacun sa chance* (*Everyone Has Their Chance*) (dir. René Pujol), released by Pathé-Natan on December 27, 1930, the first feature film with Jean Gabin as lead actor. It was produced in the Pathé studio of Joinville-le-Pont, in coproduction with the German Marcel Hellmann Film (a German version, *Kopfüber ins Glück*, dir. Hans Steinhoff, was produced simultaneously). The film is introduced by one of the actors, André Urban, as if he was on stage speaking to a theatre audience. Then, each of the actors appears in front of the camera to bow when their names are called.

Figure 8.1: Sound engineer Antoine Archimbaud in *Le mystère de la chambre jaune* (*The Mystery of the Yellow Room*) (1930, dir. Marcel L'Herbier).

so prominently in the opening credits. Then, the voice reads the names of the camera crew, and we see the big, blimped cameras, which makes this sequence essentially a short documentary about the newest technologies of the day. Then comes the sound editing by Jacques Manuel and Marguerite Baugé, and we see their hands on the soundtracks. It is only then that the actors are introduced. Thus, the credits function to mentally prepare the audience through sound and image: this will be a film full of suspense where the sound will be fundamental. The credit sequence ends with the film title, rendered by letters of fire and accompanied by sounds of wind, like in a horror movie.

The first scene after the credits employs bold camera angles to show servants denying journalists entry to a ceremony. But one camera follows someone sneaking through the crowd with a very subtle tracking shot. Obviously, this shot was filmed silently, and the sound was added later, thus allowing for very flexible camera movements. The scene continues with a 180-degree pan and a tracking shot that brings us to a majestic door. Then, we see the shadow of a stranger sitting down and quickly standing up again to the sound of the Marseillaise. The shot is one minute long, with many changes in the camera axis. While the global appreciation of *Le mystère de la chambre jaune* has focused on the crime plot, analysing these formal details reveals the great care that went into the planning of fluid camera movements and the 'fight against stasis' in this early sound film.

On November 14, 1930, *Le chemin du paradis* was released in France. Filmed in two versions in Germany by Wilhelm Thiele (*Die Drei von der Tankstelle*) and Max de Vaucorbeil, this is my only example not made at a French studio. The film was an enormous success in both countries. The first eleven shots of a car racing down a road only last between one and two seconds each, filmed from a variety of angles: at the level of the asphalt, from a low angle, a high angle, or with a mise en abyme in the rear-view mirror. The music consists of klaxon horns and then a complete orchestra. The scene with its acrobatic cinematography (including abrupt camera movements) was filmed silently, with the klaxon horns post-synchronized. Next, a static shot shows the three singing protagonists in a car, filmed against rear projection, with the car being moved by stage technicians on the set. This way, it was possible to record the most famous song of the movie ("Avoir un bon copain"/"Ein Freund, ein guter Freund") with an off-screen orchestra.

Pathé-Natan released *La petite Lise* (*Little Lise*) (dir. Jean Grémillon) on December 9, 1930. As a musician who had played music in movie theatres before becoming a director, Grémillon paid a lot of attention to the way sound was recorded and edited, and he sometimes even composed the music for his films (even though for *La petite Lise*, the composer was Roland Manuel). Already during the opening credits, we hear a strange music with a female voice singing a song about convicts. Then, a documentary-style scene takes us to the prison of Cayenne. In the courtyard, the camera follows the movements of the characters – prisoners and guards – with wide pans. Due to the direct sound, their voices sometimes sound muffled and distant. After this exterior scene, there is a very long tracking shot (more than two minutes) inside the prison, with shouts, songs, and half-naked convicts showing their tattooed bodies. We follow the hubbub of conversations, filmed with sudden camera movements. The convicts play, cook, and call out to each other. One tries to sell meatballs for three francs each: "Trois francs la boulette!" The soundtrack consists of a mix of overlapping dialogue and noises, while the camera seems to capture moments of the prisoners' lives. In synchronous live sound, we hear convicts singing a love song: "Ferme tes jolis yeux, au doux pays des rêves …" ("Close your beautiful eyes in the sweet land of dreams …"). It is said that producer Bernard Natan violently rejected the film, and little publicity was done.[14] But the critics loved the filmmaker's experimentation:

14 Sellier, *Jean Grémillon*, 14.

> The condemned go through their awful lives, crushed by fate. The scenes of the dormitory, the noisy games of the inmates, the escape of two of them are among the best. A guard comes and suddenly there is relentless silence. A 'tracking shot' allows us to see the long and sinister room on which this agonizing silence weighs. For it is one of the characteristics of La Petite Lise, M. Grémillon knows how to give all the power to silence.[15]

In total, *La petite Lise* contains eighteen minutes of astonishing sequences that take place in the prison. In these scenes, Grémillon experiments with many ways of using sound and playing with the camera, as he makes it jump from one place to another, so that the film never comes close to the infamous 'canned theatre.'

1931: Abundant Experimentation

On March 4, 1931, *David Golder* (dir. Julien Duvivier), produced by Marcel Vandal and Charles Delac, was released in France. The music of the credits, composed by Walter Goehr, begins in a post-romantic style and then turns into dance music. This is followed by a kind of montage sequence consisting of very short shots, many of which seem to be stock footage shot silently and then post-synchronized. Images of the Paris stock exchange appear in superimposition with the shot of a man audibly announcing a stock market price, mixed with the shouts of the crowd. This is followed by images that show the modern transportation of goods: a very short shot of a boat with the accompanying sound of a siren, followed by superimposed images of rails and the sound of a train. Then comes a plane with the sound of its engine. Once again, we see the turmoil of the Palais Brongniart Stock Exchange, with the shouts to buy and sell. Two men appear in superimposition, and one of them says: "Golder is a scoundrel" (Fig. 8.2). This is followed by views of a cargo ship and the sound of sirens, then, again, rails with the sound of a train engine. After this, once more, the stock exchange with its hubbub and two men in superimposition, this time accompanied by the words "Golder, my old man, he's a great guy." Next, we see the chimney of a train going full speed, and we hear the whistle of the locomotive. Finally, a match cut creates a transition from the closing of the door of a rail wagon to the closing of the door of an apartment (both audible), which marks the beginning of the first dialogue scene. Inside the apartment, the camera follows the protagonist,

15 "La Petite Lise."

Figure 8.2: Superimposition with dialogue in *David Golder* (1931, dir. Julien Duvivier).

businessman David Golder (Harry Baur), with an impressive movement, recording the conversation in direct sound.

In a sense, this opening sequence of *David Golder* sounds like Walter Ruttmann's *Weekend* from 1930, a "film without images" and a precursor of the experimental *musique concrète* of Pierre Schaeffer. One year later, to convey the hectic pace of the world of high finance in the first seconds of the film, Julien Duvivier creates a rapid succession and superimposition of images and sounds.

A few weeks later, on May 21, 1931, the independent company Braunberger-Richebé, pioneers of sound film in France who used the Western Electric system, released *Le Blanc et le Noir* (*Black and White*, dir. Robert Florey), with a screenplay adapted by Sacha Guitry from his own play. The credits insist on this relation to a stage play. We see a hand turning the pages of a book (like the printed version of a play), with the first page showing a half-open theatre curtain. A later page describes the film as "Four acts by Sacha Guitry," and the credits conclude with a page saying, "First act: in a Palace near the Spanish border." But instead of a static camera shooting the set as if it was a stage, the following sequence begins with a very long and complex tracking shot. The camera is in an elevator going down to the grand salon of a luxurious hotel. There is no doubt that Florey, who was a film critic before becoming a filmmaker, pays homage to Friedrich Wilhelm Murnau's famous tracking shot from the opening of *Der letzte Mann* (*The Last Man/The Last Laugh*) from 1924. In Murnau's film, cinematographer Karl Freund's camera went down with an elevator entering the lobby of a grand hotel in Berlin for ten seconds. But Florey decided to make the shot last even longer. His sequence shot is almost two minutes long with no cuts. The camera moves out of the elevator and then from one group to another. It moves smoothly, first showing a couple, then another, then a group of four. Conversations follow one another without mixing while the music from the opening credits continues. Among others, Fernandel in one of his first roles in a feature film, appears as a hotel employee and starts a conversation with a couple.

After the first cut to a different part of the hotel lobby, the main character is heard speaking to his friend Georges, but all we see of him is a hand gesturing out of a comfortable leather armchair. The hero is played by the famous Raimu, well known to theatregoers in Marseille, Paris, and other big cities in France, but also to film fans thanks to *Marius* (dir. Alexander Korda), a huge screen success in 1930. Raimu's voice, with a deep Marseille accent, is enough for filmgoers to know who is speaking. Then, the character is quickly *de-acousmatized*, that is, his face is shown. This foreshadows the importance of the *acousmêtre* (an acousmatic being, a person that we hear but can't see) in this film.[16] Later on, the hero's wife will have a sexual relationship with a man who is completely in the dark. We only hear the voice of this stranger, who won't be de-acousmatized. The only clue as to who he may be comes from his voice when he sings Black spirituals.

June 21, 1931 saw the release of *On purge bébé* (*Baby Gets His Purge*), Jean Renoir's adaptation of Georges Feydeau's play and his first sound film. The film demonstrates very little effort to distinguish itself from 'filmed theatre.' According to Renoir, it was prepared in one week, filmed in one week, and edited in one week, being less a 'real film' than a "summary examination"[17] for the producer to see if Jean Renoir could direct a talkie. Renoir recalled: "Three weeks after shooting began the film was shown at the Gaumont

16 The concepts of *acousmêtre* and *de-acousmatization* are explained in several books by Michel Chion (for instance, in *Audio-Vision: Sound on* Screen, 1994), but they are easier to find online, in Chion, "Glossary."

17 Renoir, *My Life and My Films*, 105–6.

Palace and brought in a lot of money."[18] In this list of 1931 French productions, it is the only film that stays close to theatre. The only 'experimental' sound in the film is that of a toilet flushing (heard from behind the door), which was recorded in a real toilet.

After *On purge bébé*, on November 20, 1931, Braunberger-Richebé released Renoir's second sound film, *La chienne* (*Isn't Life a Bitch?*). In stark contrast to *On purge bébé*, the camera moves a lot here, and the acting style is far from theatrical. As if Jean Renoir was mocking the idea of 'filmed theatre,' we first see and hear puppets in a puppet theatre explaining what kind of play we are about see. The lines spoken by the puppets overlap, establishing a French tradition of experimental dialogue where it's difficult to understand what is said because several voices are heard at the same time.[19] After this introduction by puppets, the camera, positioned in a restaurant dumbwaiter, shows the point of view of the food. We stay in this narrow lift, overhearing the conversation during a dinner among office colleagues. Then, a tracking shot along the dinner guests shows conversation and laughter, which contrasts with the sad character of Legrand (Michel Simon). Lively music and a lewd double entendre by his colleagues do not change his mood. But outside the restaurant, Legrand intervenes to protect a battered woman. With direct, synchronous sound recorded on location in the streets of Paris, Renoir uses the noises of the city to create a feeling of proximity between his main characters – the prostitute, the pimp, and the naïve client.

My last example is *Les cinq gentlemen maudits* (*The Five Cursed Gentlemen*) (dir. Julien Duvivier). Produced in late 1931 by Les Films Lutetia, the film was released on June 7, 1932. At the beginning of the credits, we hear a Moroccan song chanted rhythmically. Then, the music of Jacques Ibert introduces different themes. We come back to the work song of Moroccans, who are tanning leather in basins in Marrakech. These are documentary sounds and images showing real workers in the Moroccan city. Then we hear the beginning of a Muslim prayer. A back tracking shot filmed from above reveals the tanneries of Marrakech behind the praying man. Later, the sound of an ocean liner mixes with the sound of the sea. The film was mostly shot on location, in Morocco and on a cruise liner. Once again, we have a film that is far away from the 'filmed theatre' shot on a studio set.

18 Ibid., 107.

19 O'Brien, *Cinema's Conversion to Sound*, 121–22.

Conclusion

In 1932, René Clair lamented: "The first talking pictures – *Melodie der Welt*, *Broadway Melody*, for instance – contained more innovations than we have ever seen in the entire production that followed. Since then, by industrial routine, by lack of boldness, we have imprisoned all cinema in the rules of filmed theatre, which should have been only a part of it."[20] The French filmmaker seems to regret the lack of on-location shooting (as in *Melodie der Welt*) and spectacular camera movements (as in *Broadway Melody*) in more recent films. In the opening sequence of *Broadway Melody*, we see a view from a plane flying over New York, with a mix of music and city noises (klaxon horns, sheep sirens). Then we hear a cacophony of songs, rehearsals, and business talk outside and inside a music publishing company. The scene cuts between distinct songs, heard in insulated booths, and the general maelstrom of the open space offices. It is probably one of the very few sequences in American cinema of the period without clearly audible dialogue – and it only lasts about one minute. By contrast, when we listen to the French films produced during the 1930s and later, we hear many instances of indistinct sounds, hubbub, and noises mixed in an original way. Moreover, as the examples in this chapter have shown, we can already see many camera movements and experimental uses of sound in films from the early 1930s, especially in their opening sequences. Even if the following sequences sometimes include more canned-theatre-style shots, experimentation would often continue throughout the entire film. Of course, a detailed look at the more than 100 French films produced annually during the period is beyond the scope of this chapter. However, the twelve examples chosen here are very representative of the different types of films made in the 1930s, coming from different studios, from famous filmmakers like René Clair, Julien Duvivier, and Jean Renoir, as well as from more 'forgotten' ones like Pierre Colombier, Robert Florey, René Pujol, Wilhelm Thiele, and Max de Vaucorbeil.

In this chapter, I decided not to address films of the avant-garde such as Jean Cocteau's *Le sang d'un poète* (*The Blood of a Poet*) or Luis Buñuel's *L'Âge d'or* (*The Golden Age*), both from the same 1930–1931 season. However,

20 Clair, "Du théâtre au cinéma," 270. Clair refers to *Melodie der Welt* (*Melody of the World*) (1929, dir. Walter Ruttmann), a German experimental documentary filmed around the world in 1928 that contains almost no dialogue, and *Broadway Melody* (dir. Harry Beaumont), produced by MGM in 1928 and released in early 1929. For more of Clair's references to *Broadway Melody*, see Clair, *Cinema Yesterday and Today*.

as I have shown, even when we look at and listen to *popular* French films from the period, we can discover numerous experiments in the majority of productions. Most mainstream productions from 1930 and 1931 contain scenes shot on location, and they experiment with sound effects such as post-synchronized voices, background voices, and unusual noises (such as the sound of a toilet flush).

Many more experimental approaches were introduced in the following years. Duvivier's *La tête d'un homme* (*A Man's Head*) (1933) plays with songs and direct sound, as does the opening sequence of this almost totally forgotten film. Some opening sequences from 1934/1935 (for instance, *Tartarin de Tarascon*, dir. Raymond Bernard; and *Justin de Marseille*, dir. Maurice Tourneur) suggest that experimentation is slowing down, and from the late 1930s on, a more classical way of using sound becomes common: music that continues from the credits into the first scene, followed by dialogue and sound effects.

More films from this period remain to be analysed in detail for the way they are produced and for their remarkable innovations. In addition, it would be interesting to compare French films with productions from other countries to see if the same kind of experimentation can be found. In any case, we can say that René Clair was wrong in 1932, when he believed that the French cinema had no more imagination after 1929.

Bibliography

Aveline, Claude. "Présentation – Muet sonore ou parlant? La Dernière compagnie – Trader Horn." *Chroniques d'un cinéphile*, 17–20. Paris: Séguier, 1994.

Barnier, Martin. *En route vers le parlant: Histoire d'une évolution technologique, économique et esthétique du cinéma (1926–1934)*. Liège: CEFAL, 2002.

Bennet, Christophe. *La musique à la Radio dans les années trente*. Paris: L'Harmattan, 2010.

Bennet, Christophe. "Pianist Kartun and Composer Erwin: From Celebrity to Their Internment." Accessed March 5, 2023, https://halshs.archives-ouvertes.fr/halshs-0114657.

Billard, Pierre. *L'Âge classique du cinéma français: Du cinéma parlant à la Nouvelle Vague*. Paris: Flammarion, 1995.

Chion, Michel. *Audio-Vision: Sound on the Screen*. New York: Columbia University Press, 1994.

Chion, Michel. *Un art sonore, le cinema: Histoire, esthétique, poétique*. Paris: Cahiers du cinéma, 2003.

Chion, Michel. "Glossary: 100 Concepts to Think and Describe Sound Cinema." Accessed February 2, 2023. http://michelchion.com/texts.
"Chiqué." *Cinémonde*, February 13, 1930.
Clair, René. "Du théâtre au cinéma." *Anthologie du cinéma*, edited by Marcel Lapierre, 264–70. Paris: La Nouvelle Édition, 1946.
Clair, René. *Cinema Yesterday and Today*. Edited by R. C. Dale. New York: Dover Publication, 1972.
Girard, René. "Sous les toits de Paris." *Cinémonde*, May 8, 1930.
Henri-Robert, Jacques. "Le film sonore. Ce que le public en pense." *Cinéa*, April 15, 1929.
"La Petite Lise." *Cinémonde*, December 11, 1930.
L'Herbier, Marcel. "Opinioni di artisti francesi sui films parlanti." *La Rivista cinematografica*, March 30, 1929.
Marie, Michel. "*Sous les toits de Paris*, film de René Clair." Accessed January 3, 2023. https://www.universalis.fr/encyclopedie/sous-les-toits-de-paris/.
Mirbel, Jean de. "Prix de beauté sur la plage de Joinville." *Cinémagazine*, September 20, 1929.
Mitry, Jean. "Entretien." *Cinématographe*, no. 47 (May 1979): 18–23.
O'Brien, Charles. *Cinema's Conversion to Sound: Technology and Film Style in France and the U.S.* Bloomington: Indiana University Press, 2005.
Renoir, Jean. *My Life and My Films*. Translated by Norman Denny. London: Collins, 1974.
Sellier, Geneviève. *Jean Grémillon: Le cinéma est à vous*. Paris: Klincksieck, 2012.
Vial, Georges. "Le danger du cinéma parlant." *Le Courrier cinématoraphique*, July 21, 1928.

About the Author

Martin Barnier is Professor of Film Studies at University of Lyon 2. He is the author of *En route vers le parlant: Histoire d'une evolution technologique, économique et esthétique du cinema (1926–1934)* (2002), *Des films français made in Hollywood: Les versions multiples (1929–1935)* (2004), *Bruits, cris, musiques de films* (2010), *Le Cinéma 3-D. Histoire, économie, technique, esthétique* (written with Kira Kitsopanidou, 2015), and *Une brève histoire du cinéma* (written with Laurent Jullier, 2nd edition 2021). He is co-editor of the issue *Le Cinéma européen et les langues* of the online journal *Mise au Point* (2013).

9. Reporters, Radio Waves, and the Dispersed Audience

Staging the Radio in Early German Sound Cinema

Jörg Schweinitz

Abstract: Early sound film was linked to the broadcasting and recording industries both economically and technologically. This alliance produced a variety of interactions between the three media, which were regarded as attractions of modernity around 1930. The widespread fascination with radio not only combined theories on film and radio but also became the subject of numerous feature films. Against the backdrop of these discourses, cinema soon developed conventional narratives about radio, which, in turn, shaped the media fantasies of the time. Based on the analysis of selected films, this chapter explores how the attraction of radio was formulated, disseminated, appropriated, and transferred to early sound films. The focus is on the (film) character of the radio reporter and the new dispersed audience united by the 'radio waves.'

Keywords: broadcasting, audio culture, intermediality, modernity, stars

Before the operetta finale in the sound film *Die große Sehnsucht* (*The Great Desire*) (1930, dir. István Székely), a technician appears on screen and carefully sets up a microphone in the film studio. He is followed by a radio reporter: Alfred Braun, a popular real-life radio man of the era in a cameo. Appearing as a character under his own name, he is supposed to report on the shooting of the film within the film that bears the same title, *Die große Sehnsucht*, as the film we see, and which metaleptically merges with it in the final musical number. After a brief conversation between the reporter and the technician as to whether the microphone is in the best position, a young woman appears. She introduces her autograph request by complimenting

Wiegand, D. (ed.), *Aesthetics of Early Sound Film: Media Change around 1930*.
Taylor & Francis Group, 2023
DOI 10.5117/9789463727372_CH09

Braun on how wonderfully he speaks. The fact that she accidentally swaps the star postcard she brought for one of Harry Liedtke leaves no doubt: the broadcaster himself is presented as the star here. Similarly, Betty Amann, Lil Dagover, Jenny Jugo, Fritz Kortner, Anny Ondra, Luis Trenker, Conrad Veidt, and numerous other popular movie stars (among them Harry Liedtke) briefly appear in the diegetic studio world 'as themselves' in this auto-thematic feature film.[1]

Before Braun can finally step up to the microphone, there is another slight delay: the production manager (Paul Henckels) turns to the newspapermen gathered in a corner of the studio, clearly not as spotlighted as Braun; he announces that lead actress Eva van Loe (Camilla Horn) is running late, and thus the recording is delayed. However, the director of the film within the film (Theodor Loos) immediately signals to the radio reporter from an elevated camera platform that he may begin. Now, Braun speaks into the microphone with the gestures and excessively clear articulation of the time: "Attention, attention, ladies and gentlemen, you are now listening to the first ever report from a sound film studio. We are broadcasting to you the shooting of the final scene of the sound film *Die große Sehnsucht*." Then we hear the acoustic signal for "Silence, please!" followed by the director's "Attention, recording!" and the big show number begins: "Ich wünsch' mir was" ("I wish for something"), a sound film hit by Friedrich Hollaender.

This sequence from the first feature-length film by Hungarian-born director István Székely, who went by Stefan Szekely in Germany (and later Steve Sekely in the USA), symbolically spotlights the era's reciprocal connections between sound film and radio. It testifies to how the talkies, which had been pushing into German cinemas since mid-1929, and radio, whose triumphant advance gained momentum in the mid-1920s, had long shared – and interacted – in the widespread enthusiasm for the period's new audio culture and its technology. The movie also shows how early sound film exploited this new facet of fascination with modernity for its own purposes. Sound film relied on showcasing itself as ultra-modern by emphasizing its own new audio technology – which it partially shared with radio – as well as radio's unique technology.

1 In regard to popularity, Braun is also brought close to Harry Liedtke in the sound film *Das Lied ist aus* (*The Song Is Ended*) (1930, dir. Géza von Bolváry) without being involved in this film. The film's hit song "Wenn das Wörtchen 'wenn' nicht wär" ("If the little word 'if' wasn't there") contains the lines: "Wenn ich wie der Liedtke beliebt wär' bei Frau'n oder funken könnt wie Alfred Braun ..." ("If I were as popular with women as Liedtke or could broadcast like Alfred Braun ...").

The radio world as part of the diegetic world of film also opened up new opportunities for cinematic storytelling. Radio as staged in film became part of the imagery of modernity, and film was involved in developing the respective visual signs. The brand-new genre of radio reportage, which presented the novelty (addressed in *Die große Sehnsucht*) of live reports from extraordinary places and events, offered a special fascination. And Alfred Braun, pioneer of the new radio genre, played an outstanding role in this. Due to his popularity, he repeatedly appeared as a radio reporter under his own name – the radio reporter par excellence – in feature films during the first three years of the talkies.

In what follows, I will explore some of these aspects of the 'radio in early sound cinema' phenomenon, including the character of the reporter, by looking at a series of films from the last years of the Weimar Republic. They will be analysed in the context of intermedial developments and discourses surrounding radio during this period.

Radio – Enthusiasm for a New Popular Medium

While the general enthusiasm for the new audio culture of radio, talkies, and phonograph records reached a peak around 1930, it has a long and well-documented prehistory going back to the nineteenth century.[2] There is also a long tradition of coupling film with separate sound recordings, which has been thoroughly researched. In Germany, Oskar Messter's Tonbilder ("sound pictures"), which circulated for a few years starting in 1903, are the most prominent example.[3] But audio culture received a tremendous boost with the establishment of radio – in Germany, with the first regular broadcasts of Berlin's radio station Funk-Stunde beginning in the fall of 1923. While Funk-Stunde had only a handful of listeners at first, the Reichspost, which operated the emerging German broadcasting network, counted more than one million radio receivers on January 1, 1926. This figure alone testifies to the enormous momentum and rapidly growing popularity of the new medium, and the trend continued to accelerate: in January 1929, there were already over 2.6 million registered radio subscribers, in 1931, the number

2 What I call "audio culture" here has been studied with regard to the American context by Donald Crafton, who uses the concept of early sound film's "electric affinities." Crafton, *The Talkies*, 23–61; see also Wurtzler, *Electric Sounds*; Lastra, *Sound Technology and the American Cinema*.

3 Koerber, "Oskar Messter, Film Pioneer," 55–57; Simeon, "Messter und die Musik des frühen Kinos," 135–47.

was 3.5 million, and in 1932 – despite the world economic crisis – almost 4 million.[4]

From the late 1920s on, record players could be connected to the radio's loudspeaker, which also gave additional impetus to the record industry. In general, the audio equipment industry experienced a tremendous boom during the Weimar years, and soon, the new technological listening culture was emblematically reflected and celebrated in major events and large-scale buildings. In line with the industry's growing self-confidence, these were conceived and perceived as landmarks of modern life: The Great German Radio Exhibition, a trade fair for the audio equipment industry as well as for broadcasters, was first held in Berlin in 1924 and would continue as an annual event. In 1926, the Berlin Radio Tower (Berliner Funkturm) was inaugurated on the exhibition grounds. And in 1929–1931, architect Hans Poelzig, an exponent of Neue Sachlichkeit (New Objectivity), oversaw the construction of the still impressive House of Broadcasting (Haus des Rundfunks) near the radio tower. By 1931, Berlin had thus created a 'radio city,' which attracted wide public interest.

Developments that were perceived as spectacular, culture-changing, and astonishing also happened in the realm of reception. For example, it became possible to receive broadcasts across borders at a very early stage. This sparked contemporaries' fascination with the "Pan-Europe of broadcasters" and its "'little border traffic' in the air," as publicist René Schickele, who lived in southwestern Germany, put it in 1931, still full of enthusiasm for the novelty: "you only have to turn the knob a little to hear Stuttgart as well as Strasbourg, Paris as well as Berlin, and 'Radio Milano' announces itself just as clearly as 'Süddeutscher Rundfunk' here – just to stay in the neighborhood."[5] Accordingly, progressive authors such as Ernst Toller in 1930 saw it as the task of the new medium (after the devastation of World War I) to speak out against national chauvinism and the "hate-producing catchphrase."[6]

In addition, radio began to network internationally, producing the first intercontinental live broadcasts. Not only the masses but intellectuals, too, were obsessed with the new popular medium and engaged in an extensive discourse. In 1931, for example, Rudolf Leonhard wrote a hymn-like poem dedicated to the loudspeaker that sounds "far around the globe."[7] Utopias

4 Braun, *Achtung, Achtung, Hier ist Berlin!* 82–86.

5 Schickele, "Paneuropa der Sender," 140–41. Unless noted otherwise, translations are my own.

6 Toller, "Internationaler Rundfunk," 135.

7 Leonhard, "Lautsprecher," 56.

of a peaceful global listening culture circulated: *Hallo! Hier Welle Erdball!* (*Hello! This is Wavelength Earth!*) was the euphoric title that radio pioneer Fritz Walter Bischoff gave to his famous radio play, which premiered in 1928.[8]

There were, however, also voices of cultural pessimism – such as Egon Friedell in 1931 – conjuring up the worst:

> The bioscope kills the human gesture only, but the sound-film the human voice as well. Radio does the same. At the same time it frees us from the obligation to concentrate, and it is now possible to enjoy Mozart and sauerkraut, the Sunday sermon and bridge. [...] The human voice has achieved omnipresence, the human gesture eternity, but at the cost of the soul. [...] We already have nightingale concerts and Papal speeches broadcast by radio. This is the decline of the occident.[9]

The extent to which talkies and radio jointly belonged to the new audio culture in the eyes of contemporaries is also evident in Friedell's unison rejection of both media.

The non-commercial nature of German broadcasting meant that, unlike in the USA, there could be no capital-based direct merger with the private-sector industries of talkies and records. Nevertheless, there were diverse intermedial relationships. The exploitation of hit songs, for example, in the triangle of sound cinema, radio, and the record industry quickly became a common and much-discussed practice in Germany as well. Ernst Jolowicz's 1932 book on radio, for example, begins with the chapter "Cinema – Gramophone – Radio" as a matter of course.[10]

Unsurprisingly, sound films were also of interest in radio programming from the very beginning. In 1929, with cinema's incipient transition to sound, Berlin's Funk-Stunde even featured radio broadcasts of the soundtracks of complete film programmes, including newsreels, briefly introduced by an announcer. Shortly afterwards, the same station broadcast the German-language version of E. A. Dupont's *Atlantik* (*Atlantic*) (1929) from Berlin's Gloria-Palast. The sound was picked up directly from the sound equipment in the cinema and transmitted to the broadcaster via a telephone line, while a reporter in the theatre simultaneously explained the film's images and action via a second line. Something similar happened at other radio

8 Bischoff, "Hallo! Hier Welle Erdball!" 124.

9 Friedell, *A Cultural History of the Modern Age*, 475.

10 Jolowicz, *Der Rundfunk*, 7–18.

stations,[11] and soon, there were regular broadcasts devoted to criticism or discussion of current film offerings.[12]

Conversely, film critics and publicists were also interested in the radio, and they sought to apply some of the theorems of early film theory to it, especially the Laocoön concept of media specificity. As early as 1924, Béla Balázs had proposed the theorization of radio in terms of a conception of art that complemented film theory: "The invention of cinematography brought us the art of film, the presentation of life solely for the eye. Radio is bringing us the new, exclusively audible art of acoustic drama, the presentation of life solely for the ear."[13] And he continues: "In truth art requires us to switch off certain senses; attempts at the *Gesamtkunstwerk* have always proven to be dilettantish, discouraging affairs."[14] Therefore, Balázs argues, it is important for the radio drama to present a story clearly and excitingly by means of sounds and noises alone. This idea would continue to shape the discourses on radio drama and reportage around 1930 and provide a theoretical basis for the widespread skepticism about sound films.[15] As late as 1933, Rudolf Arnheim argued quite similarly in the manuscript for his book on radio.[16]

Staging the Astonishment at the New Medium

When early sound films take radio as their theme, they tie in with the general cultural enthusiasm about the medium. The fascination that emanated from the new broadcasting technology alone is apparent from Rudolf Arnheim's aforementioned book on radio. In his introduction, he describes the "many extraordinary sensations" that he experienced during a visit to the new House of Broadcasting in Berlin:

> The carpeted rooms where no footstep sounds and whose walls deaden the voice, the countless doors and corridors with their bright little light-signals, the mystifying ceremonial of the actors in their shirt-sleeves who, as if attracted and repelled by the microphone, alternately approach and withdraw from the surgical charms of the metal stands;

11 Mühl-Benninghaus, *Das Ringen um den Tonfilm*, 252.

12 For details on the beginnings of film criticism in German broadcasting from 1930 on, ibid., 256–59.

13 Balázs, "Radio Drama," 47.

14 Ibid., 48.

15 Galili, "Intermedial Thought in Classical Film Theory," 391–99.

16 Arnheim, *Radio*, 133–203. Note that the original German typescript dates to 1933.

> whose performance can be watched through a pane of glass far away as in an aquarium, while their voices come strange and near from the control-loudspeaker in the listening room; the serious young man at the controlboard who with his black knobs turns voices and sounds off and on like a stream of water [...].[17]

The palpable astonishment at the technological sensations that Arnheim expresses here is a recurring theme in early sound films dedicated to radio. They participated in the staging of the new sound and radio technology, if only to exhibit their own modernity. Thus, it seems only logical that *Die große Sehnsucht* (which ends with Braun's radio reportage) begins in a sound engineering room crammed with ultra-modern equipment. While this is not explicitly the technology of radio but rather that of a (diegetic) film studio, it does not differ fundamentally from the corresponding sound equipment in a radio studio, neither visually nor in terms of its novelty. It was only in the fall of 1929 that Ufa had put into operation its new "Tonkreuz" ("sound cross)" – Germany's first purpose-built sound film studio building, which was much admired by the press – at the Babelsberg studio. With close-ups of ornamental-looking apparatus details, *Die große Sehnsucht* stages a technoid visual world. This affinity for technology is closely related to the delight that New Objectivity took in rotating pistons and flywheels, in ornamental steel constructions and the like, as expressed in *Das Lied vom Leben* (*Song of Life*) (1931, dir. Alexander Granowsky). But instead of the visual appeal of the mechanical it is the new electrical engineering of sound that is confidently exhibited as a visual sensation here. Images like these recur quite often, for instance in *Spione im Savoy-Hotel* (*The Gala Performance*) (1932, dir. Friedrich Zelnik), another film that begins with images of the sound and broadcasting technology of radio (a sequence I will examine in more detail later).

In addition to the new radio technology, the aspect of reception is also presented as an attraction. Around 1930, radio listening finally began to become a part of modern everyday life, but it was still new enough not to be taken for granted, as can be seen in some essays by intellectuals. Radio listening still attracted attention as a new, rapidly evolving cultural phenomenon. Thus, it appears especially in urban comedies like *Wer nimmt die Liebe ernst?* (*Who Takes Love Seriously?*) (1931, dir. Erich Engel)[18] or *Ich*

17 Ibid., 19–20.

18 I thank Selina Hangartner (Zurich) for pointing out this film. She is currently working at her PhD thesis on *Reflexivity and Irony in Early German Sound Film*.

bei Tag und Du bei Nacht (*I by Day, You by Night*) (1932, dir. Ludwig Berger), where it is presented as an aspect of modern everyday life. In Berger's film, for example, it fits in well with the mechanized life of the young protagonist (Käthe von Nagy): during the day, she works in a beauty salon with machines that seem grotesque today but were ultramodern at the time, and at home, she listens to the radio in bed at night. Listening to the radio not only helps to characterize the young urban woman, it also provides a diegetic motivation for repeating the film's hit song, "Wenn ich sonntags in mein Kino geh'" ("When I go to the movies on Sundays") by the Comedian Harmonists. This was a common strategy in musical films of those years. On the one hand, they repeated their main hit several times, almost like a leitmotif, and on the other hand, it generally seemed important to diegetically justify singing.

Aside from operetta films such as *Die Drei von der Tankstelle* (*Three from the Filling Station*) (1930, dir. Wilhelm Thiele) or *Der Kongreß tanzt* (*The Congress Dances*, 1931, dir. Erik Charell), which borrowed directly from stage operetta conventions, hardly any German sound film from 1929–1932 featured vocal numbers that were not diegetically motivated somehow.[19] Therefore, as Brigitte Mayr also observes, "themes were contrived that had a musical milieu"[20] suitable for revues, ballets, and songs. The presence of a radio (or a gramophone) provided an additional option, especially for repetitions. The need for diegetic motivation is also reflected in film theory. To once again quote Rudolf Arnheim, here in an essay from 1934:

> The great difficulty is that we humans normally find ourselves in musical situations much less often than is desirable for the sound film. [...] they looked for material in which musicians played an extensive role. [...] Love stories, otherwise reserved for mute and unmusical bon vivants, were now experienced by singers: the singer undertook car chases, attended costume balls, jumped off the express train, [...] dueled and dressed up, and yet still took the time to deliver one of the multi-verse songs every few minutes [...].[21]

Similarly motivated by realism, the motif of radio provided diegetic bridges between different narrative spaces. These made it possible to accelerate the narration, to make it more complex, and to exhibit the medium itself as an

19 On the use of nondiegetic music in early German sound films see the article by Daniel Wiegand in this volume.

20 Mayr, "Budapester Charme und Wiener Ambiente," 169.

21 Arnheim, "Die Zukunft des Tonfilms," 27–28.

attraction. Reportages, live broadcasts, or radio news within the film plot played a special role here. As narrative devices, they relayed information about events that were happening elsewhere to film viewers, who were simultaneously able to hear and see how the characters received this same information.

An example can be found in a slightly later urban comedy, made in Hungary by István Székely, *Lovagias Ügy* (*An Affair of Honor*) (1937).[22] While the daughter of a strict father has to stay home, she is able to follow the live broadcast of a 'radio concert' in the Paradise Bar, where her beloved is performing as a singer. And he addresses her directly from the stage via the radio microphone, declaring his love. The diegetic bridge between the spaces – and thus the film's happy ending – is then symbolically exhibited through a trick shot, a kiss over the radio: while the young woman approaches the radio (shown from the side) as in a kissing scene, the singer appears inside the radio's dark case in double exposure, moving toward her until a symbolic union occurs in the image – the perfect visualization of the diegetic bridge.

Simpler forms of diegetic bridges via radio (without such symbolic visualization) are also common in European films of that time. For example, the French film *Le chanteur inconnu* (*The Unknown Singer*) (1931, dir. Victor Tourjansky), who otherwise worked mainly in Germany, is based on the title's mysterious singer being recognized on the radio by a character at home.

In other cases, films present the intercontinental bridge made possible by radio as a phenomenon of modernity, as *L'Inhumaine* (*The Inhuman Woman*) (1924, dir. Marcel L'Herbier) had already done. The Austrian production *Ein Sonnenstrahl* (*Ray of Sunshine*) (1933, dir. Paul Fejos) highlights the radio, which created transnational listening spaces for the first time, as part of the new consumer culture: an unemployed couple has taken a temporary job as cleaners, working alone at night in a Viennese department store. They are attracted by a beach display whose consumer world symbolizes to them the good life they have been dreaming of. They cannot resist and dress up in the beach wear on display, using the deck chairs, while the model of an ocean liner is passing by in the background. They turn on the radio in the exhibit and, in a kind of game, participate in the dance party of an imaginary American millionaire in Miami, Florida, which the radio broadcasts live.

22 Jan-Christopher Horak considers films like this one, "made by returning Hungarian remigrants who had worked in Germany before 1933," as "exile films"; he specifically names István Székely. Horak, "Exilfilm 1933–1945," 104.

Another kind of bridging function that attracted attention in discourses of the time was the technical bridge between broadcaster and listener as well as between listeners. After all, radio had not just created a new form of connection between audience and performer but also a new audience, about whose peculiarity Rudolf Leonhardt reflected in 1924: "Let us imagine: he [the listener] sits in his room, with headphones or in front of the loudspeaker, he is alone – and connected with thousands over the wave."[23] The members of this new audience simultaneously share the same listening experience, and they know that they are listening at the same time as others. Yet they also remain anonymous and spatially separated from each other, as well as from the broadcaster. In addition, this historically new type of audience, which is now referred to as a "dispersed audience" (*disperses Publikum*, a term coined by Gerhard Maletzke),[24] incorporated a cross-section of society in the era of classical radio.

Early on, sound film developed a visual structure for the narrative presentation of this kind of audience, which quickly established itself. The beginning of *Das Lied einer Nacht* (*Tell Me Tonight*) (1932, dir. Anatole Litvak) can serve as a prototypical example. Like *Das lockende Ziel* (*The Alluring Goal*) (1930, dir. Max Reichmann), a film revolving around star tenor Richard Tauber and based almost entirely on song, Litvak's film also centred on a star tenor (Jan Kiepura), and it is not by chance that both films feature a radio sequence. In *Das Lied einer Nacht*, it serves as the very beginning. Its structure is interesting: it starts with a shot of a speaker (visible as a silhouette) in a radio studio (Fig. 9.1), dissolved into an upward panning shot of the broadcast antenna and then a downward panning shot of the receiving antenna on the roof of an apartment building. Next, we see a series of short scenes showing a variety of radio listeners in different rooms. Completely absorbed by listening, they all make mistakes in various situations: an elderly gentleman with a cigar, absorbed in the radio programme, does not notice the burglar who is clearing out the safe behind his armchair; a secretary types the chorus lyrics of a hit song playing on the office radio into a business letter. The charm of the sequence lies in the comedic play with these blunders and in the way the editing purposefully misleads viewers (an extreme close-up of an open mouth, for instance, turns out to belong to a patient at the dentist's rather than to the singer heard on the radio, as one would expect). In terms of media history, however, this sequence is remarkable for another reason: it establishes a cinematic-narrative structure to show

23 Leonhardt, "Technik und Kunstform," 71–72.

24 Maletzke, *Psychologie der Massenkommunikation*, 30.

Figure 9.1: Silhouette of a radio speaker at the beginning of *Das Lied einer Nacht* (*Tell Me Tonight*) (1932, dir. Anatole Litvak).

the connection between the broadcaster and a dispersed, cross-sectional audience – a structure that was hinted at in Tourjansky's earlier *Le chanteur inconnu* (also in the opening) and that would be used frequently thereafter.

The Radio Reporter as Transmedial Star: *Spies at the Savoy Hotel*

Spione im Savoy-Hotel, literally translated: "Spies at the Savoy Hotel," which premiered in Berlin in November 1932, encapsulates much of what has been discussed so far about the staging of radio in early sound films. Moreover, it is a film focusing on the aforementioned radio genre of live reportage – and with it, once again, on the star reporter of Funk-Stunde, Alfred Braun. As in *Die Große Sehnsucht* from 1930, Braun appears 'as himself,' here in an even more prominent role. I will thus conclude by examining the film and its hero, the 'racing radio reporter,' in a little more detail.

Like Litvak's *Das Lied einer Nacht*, which premiered in May 1932, *Spione im Savoy-Hotel* begins in the radio studio: a shot is fired, and actors appear in silhouette (this, too, is a parallel to Litvak's opening; it is obviously intended to visually express the 'blindness' of radio). Then, a (fully visible) announcer steps up to a microphone which – like all the following microphones – bears the eye-catching lettering "Funk-Stunde" on its suspension ring (a deliberate product placement, especially since the station's actual microphones bore no such lettering). The speaker proclaims the end of the crime radio play and announces the next programme after the required changeover break: "We'll be back in two minutes with the broadcast from the Savoy Hotel, with Alfred Braun at the microphone." This is followed by a montage sequence consisting of documentary shots, underscored by the acoustic station signal of Funk-Stunde: First, the camera slowly moves along the screen-filling lettering "Haus des Rundfunks" on the exterior of the emblematic broadcasting building. This pan dissolves into an image of the radio tower, Berlin's other landmark of the new medium. Then, a series of images exhibit the sound and broadcasting technology inside the building: switchboards, measuring instruments with deflecting pointers, generators, capacitors, entire batteries of imposing radio tubes, etc., are operated and shown in detail – until the station signal falls silent and Braun's voice sounds from a control loudspeaker in the studio (shown at close range): "Attention, attention, ladies and gentlemen!" There is a slow transition from the loudspeaker to Braun and his assistant Gottlieb (Erich Kestin). Both are standing in the ballroom of the fictional Savoy Hotel with their microphone (Fig. 9.2).

This beginning sets the stage in several respects: there is, first, the elaborate cinematic staging of the new radio world. The enthusiasm for the technological dimension in the spirit of New Objectivity reveals the hand of Reimar Kuntze, one of the cinematographers of Walter Ruttmann's *Berlin. Die Sinfonie der Großstadt* (*Berlin: Symphony of a Great City*) (1927). Much of the fascination that was palpable in Arnheim's description of the interiors of the same broadcasting house is communicated visually. Second, the film's protagonist, Alfred Braun, is introduced as the virtual personification of the new medium.

Braun himself immediately explains his special role in the diegetic world: "You are listening to a broadcast from the charity tea party at the Savoy Hotel. […] I am now acting in a double capacity, not only as announcer of the programme here but also as speaker of the radio transmission." Accompanied by sidekick Gottlieb, who carries the microphone, he walks through the audience rows describing the scene, as curious and spellbound hotel employees watch through door slits and windows. Finally, Braun announces the fictitious Gloria Jazz Band as well as the (really existing)

Figure 9.2: Radio reporter Alfred Braun and his assistant in *Spione im Savoy-Hotel* (*The Gala Performance*) (1932, dir. Friedrich Zelnik).

clown troupe Fratellini Brothers & Gino, who perform on stage. Braun's activities indicate a third capacity: he guides the cinema audience through the film, commenting and later also acting, partly as a provider of narrative information, partly as a central character.

The film succeeds in combining two genres, the musical and the detective film. On the one hand, the stage numbers, shown in detail and commented on for the radio listeners (and the cinema audience), are plausibly motivated through their location on the hotel's diegetic stages and later in the fictitious Tivoli variety theatre; on the other hand, a murderous detective story unfolds in this hotel environment. The reporter gets caught up in this and, as the 'man of action,' resolutely takes the investigation into his own hands. Along the way, the resolution of the crime plot also resolves the second plot line – a story of love and jealousy between Susi (Margot Walter), a young musician, and the leader of the jazz band – which Braun also takes care of on the side. Gottlieb is almost always present with the microphone, so that the imaginary radio audience can always follow the action. This, of course, showcases the celebrated attraction of the reportage as a spectacular live performance in exaggerated form.

That *Spione im Savoy-Hotel* is a star vehicle for Alfred Braun is evident from how dominant his character is – compared with both the bumbling

Gottlieb as his comic foil and the other protagonists, played by film stars of the period, such as Olga Chekhova (the spy) and Alfred Abel (the diplomat Mr. Palmer).

What was the basis of the real-life reporter's star status that the film draws on? It was certainly less the fact that Braun had been an actor and director at various Berlin theatres since 1907 and had occasionally appeared in feature films since 1916.[25] Rather, the decisive factor is that this experience brought him to Funk-Stunde as early as May 1924, only six months after the programme first went on air. Working as a programme maker, radio play director, and presenter, he quickly became popular with his distinctive voice. In addition to studio work, he began to report live from events in the city, using a mixture of original sounds and his voice for description and commentary. He has thus been credited with the 'invention' of the radio reportage (at least in Germany). The format quickly found imitators, but it testifies to Braun's special status and that of the reported events that some of his live reports were recorded on discs that have survived as audio documents. Examples include the funeral procession for Foreign Minister Gustav Stresemann through Berlin in October 1929 and the Nobel Prize award ceremony for Thomas Mann in Stockholm in December 1929. With the latter, Braun introduced the format of the so-called "Flüsterreportage" ("whispered reportage,") which underscored the intimacy and solemnity of the event. But he also reported on everyday life, for instance from industrial workshops, the newspaper printing plant at Ullsteinhaus, or a depot of Berlin's public transportation system, in each case using the original soundscape for authenticity. In the era of New Objectivity, the choice of such settings was considered impressive. Ernst Jolowicz remarked on the art of reportage: "The more can be heard of the event near the microphone, the more associations and audible illustrative materials are available to the reporter."[26]

Early on, reportage sites included cabarets and variety theatres such as Berlin's Scala or the Plaza, the sounds of which appealed to listeners' demand for entertainment (and which provided models for the settings of *Spione im Savoy-Hotel*). Braun later recalled, "soon there was a whole series of entertainment venues in Berlin with their own constant broadcast feed, to which the reporter could tune in at any time as desired."[27] In short, Alfred Braun had become the presenter of modern life, a star who helped shape the imaginative world of urban modernity.

25 Entry "Alfred Braun" on filmportal.de.

26 Jolowicz, *Der Rundfunk*, 85.

27 Braun, *Achtung, Achtung, Hier ist Berlin!*, 67.

No wonder that he was also admired by many intellectuals. Hermann Kasack, for example, praised his radio reports:

> Not only the events but above all their intensity, their atmosphere, the reflection of the effect must be transmitted as if on their own through word, sound, pause, announcement. In this, Braun is unequalled in all of Germany. He has the ability to adapt to an unknown situation in such a way that he makes audible not just what he sees and observes but above all what is worth seeing and observing for the listener at the radio![28]

The station and Braun himself further increased his fame as Braun assumed a prominent role in the public self-presentation of the new medium. In addition to in-person appearances for Funk-Stunde at Berlin's annual radio exhibitions, he was also present as a wax figure for their advertising activities, and at the festive opening of the Berlin Radio Tower in 1926, where he was chosen to deliver the ceremonial prologue.[29] Thus, it seems only logical that Braun – by now the voice as well as the face of the station – was already involved in films dealing with radio during the silent era beginning in 1926. These were shorter, sometimes semi-documentary films that promoted the new medium. Among them were the (surviving) documentary *Die Geburt des Reichsrundfunks in Berlin im Jahre 1923* (*The Birth of Reichsrundfunk in Berlin in 1923*) (probably around 1927, dir. Oskar Mamis), the probably lost semi-documentary *Der sprechende Film* (*The Talking Film*) (1926, Dir. unknown), and the farcical *Funkzauber. Ein Volksstück von der Liebe auf des Rundfunks Wellen* (*Radio Magic: A Folk Play about Love on the Radio Waves*) (1927, dir. Richard Oswald). In all these films, Braun appears as a radio announcer. By 1929/30, he expanded his intermedial presence to include records when he recorded the "Reportage-Couplet" from an operetta by Ralph Benatzky (text by Arthur Rebner) with orchestral accompaniment for the Homocord label.[30]

This background of Braun's enormous popularity accounts for his role as a radio reporter in the early talkies – and especially in *Spione im Savoy-Hotel*. He plays a character who masterfully pulls all the strings of the plot, and at the same time, he can rely on 'his' medium, which helps to see through the events, to make them plausible, and to resolve them. For example, the

28 Kasack, "Mikroreportage," 130.

29 Braun, *Achtung, Achtung, Hier ist Berlin!*, 23, 25; see also Fig. 15.

30 The original sound of the Homocord record is available on YouTube: https://www.youtube.com/watch?v=rJQJDqv1dQI. Accessed March 11, 2023.

arrival of the 'wanted spy' Countess Livskaya, alias Miss Harris, in Berlin is first announced in a radio message, thus providing the character who appears shortly thereafter with a background characterization for Braun as well as the film's audience. Braun's comments also make plausible the repetition of the film's hit song, "Mach mir's nicht so schwer" ("Don't make it so hard for me") by the Comedian Harmonists, which is performed in three versions on the diegetic stage. And finally, when a member of the Gloria Jazz Band – who was involved in the criminal action and had disguised himself as a clown – is unmasked and arrested after a backstage chase, the reportage that partially accompanied the action also facilitates understanding.

In the end, the radio microphone delivers the final proof needed to convict the spy. When Braun confronts the fake countess and reveals his knowledge as an eyewitness, she pulls out a revolver: "Apart from you, no one knows ..." The reporter remains cold-blooded: "You are mistaken, Countess, hundreds of thousands already know" – he pushes aside a curtain to reveal Gottlieb with the open microphone and the police.

This scene is reminiscent of the film *Der Schuss im Tonfilmatelier* (*The Shot in the Talker Studio*) (1930, dir. Alfred Zeisler), which is set in a film studio and whose murder story is ultimately resolved because a technician had turned his microphone on before filming, thereby recording a telltale conversation that convicts the murderer. Zeisler's film – just like *Die große Sehnsucht* discussed at the beginning – belonged to a wave of German feature films from 1930–1932, in which the sound film talks about itself and exhibits its attraction, the new sound technology.[31] Films like *Spione im Savoy-Hotel*, which present the new medium of radio, obviously follow the same strategy in order to celebrate their own modernity.

It is only fitting that the film ends with the radio reporter saying goodbye, using the usual formula for closing the programme: "Ladies and gentlemen, the show is over. Our broadcast from the Tivoli is over [...]. We wish you a good night. Please don't forget to ground the antenna."

Bibliography

Arnheim, Rudolf. *Radio* [1933]. Translated by Margaret Ludwig and Herbert Read. London: Faber & Faber, 1936.

Arnheim, Rudolf. "Die Zukunft des Tonfilms" [1934]. *Montage AV* 9.2 (2000): 18–32.

31 Schweinitz, "Wie im Kino!," 373–92.

Balázs, Béla. "Radio Drama" [1924]. Translated by Russell Stockman. *October* 115 (Winter 2006): 47–48.

Bischoff, F[ritz] Walter. "Hallo! Hier Welle Erdball!" [1928] In *Radio-Kultur in der Weimarer Republik*, edited by Irmela Schneider, 124. Tübingen: Narr, 1984.

Braun, Alfred. *Achtung, Achtung, Hier ist Berlin! Aus der Geschichte des deutschen Rundfunks in Berlin 1923–1932*. Berlin: Haude & Spener, 1968.

Crafton, Donald. *The Talkies: American Cinema's Transition to Sound, 1926–1931*. Berkeley, Los Angeles, London: University of California Press, 1997.

Friedell, Egon. *A Cultural History of the Modern Age: The Crisis of the European Soul from the Black Death to the World War* [1927–1931]. Translated by Charles F. Atkinson. New York: Knopf, 1930–1932.

Galili, Doron. "Intermedial Thought in Classical Film Theory: Balázs, Arnheim, and Benjamin on Film and Radio." *The Germanic Review* 88, no. 4 (2013): 391–99.

Horak, Jan-Christopher. "Exilfilm 1933–1945." In *Geschichte des deutschen Films*, edited by Wolfgang Jacobsen, Anton Kaes, and Hans Helmut Prinzler, 101–18. Stuttgart: Metzler, 1993.

Jolowicz, Ernst. *Der Rundfunk. Eine psychologische Untersuchung*. Berlin: Max Hesse, 1932.

Kasack, Hermann. "Mikroreportage." In *Radio-Kultur in der Weimarer Republik*, edited by Irmela Schneider, 128–33. Tübingen: Narr, 1984.

Koerber, Martin. "Oskar Messter, Film Pioneer: Early Cinema between Science, Spectacle, and Commerce." In *A Second Life: German Cinema's First Decades*, edited by Thomas Elsaesser and Michael Wedel, 51–61. Amsterdam: Amsterdam University Press, 1996.

Lastra, James. *Sound Technology and the American Cinema: Perception, Representation, Modernity*. New York: Columbia University Press, 2000.

Leonhard, Rudolf. "Lautsprecher" [1931]. In *Radio-Kultur in der Weimarer Republik*, edited by Irmela Schneider, 56. Tübingen: Narr, 1984.

Leonhard, Rudolf. "Technik und Kunstform" [1924]. In *Radio-Kultur in der Weimarer Republik*, edited by Irmela Schneider 67–72. Tübingen: Narr, 1984.

Maletzke, Gerhard. *Psychologie der Massenkommunikation. Theorie und Systematik*. Hamburg: Hans Bredow Institut, 1963.

Mayr, Brigitte. "Budapester Charme und Wiener Ambiente. Der Musikfilm im Exil in Österreich und Ungarn 1933–1938." In *Als die Filme singen lernten. Innovation und Tradition im Musikfilm 1928–1938*, edited by Malte Hagener and Jan Hans, 166–78. München: edition text+kritik, 1999.

Mühl-Benninghaus, Wolfgang. *Das Ringen um den Tonfilm. Strategien der Elektro- und der Filmindustrie in den 20er und 30er Jahren*. Düsseldorf: Droste, 1999.

Schickele, René. "Paneuropa der Sender" [1931]. In *Radio-Kultur in der Weimarer Republik*, edited by Irmela Schneider, 140–43. Tübingen: Narr, 1984.

Schweinitz, Jörg. "Wie im Kino! Die autothematische Welle im frühen Tonfilm. Figurationen des Selbstreflexiven." In: *Diesseits der 'Dämonischen Leinwand': Neue Perspektiven auf das späte Weimarer Kino*, edited by Thomas Koebner, 373–92. München: edition text + kritik, 2003.

Simeon, Ennio. "Messter und die Musik des frühen Kinos." In *Oskar Messter. Filmpionier der Kaiserzeit,* edited by Martin Loiperdinger, 135–47. Basel, Frankfurt am Main: Stroemfeld, 1994.

Toller, Ernst. "Internationaler Rundfunk" [1930]. In *Radio-Kultur in der Weimarer Republik,* edited by Irmela Schneider, 134–35. Tübingen: Narr, 1984.

Wurtzler, Steve. *Electric Sounds: Technological Change and the Rise of Corporate Mass Media.* New York: Columbia University Press, 2007.

About the Author

Jörg Schweinitz is retired Professor of Film Studies at University of Zurich. He is the author of *Film and Stereotype: A Challenge for Cinema and Theory* (Columbia University Press 2011) and editor of several articles and books on film history, including *Film Bild Kunst: Visuelle Ästhetik des vorklassischen Stummfilms* (edited with Daniel Wiegand, 2016) and *Die Zeit des Bildes ist angebrochen: Französische Intellektuelle, Künstler und Filmkritiker über das Kino, 1906–1929* (edited with Margrit Tröhler, 2016). He is co-editor of *Montage AV.*

IV

Speech and Language

10. Talking Photographs

The Speaking Subject in Anglophone Newsreel and Documentary (1927–1936)

Irina Leimbacher

Abstract: While feature films and newsreels almost immediately exploited the use of direct sound, filmmakers working in the field of documentary focused their creative energies on other aspects of cinematic soundscapes. Orchestrated noises, music, and occasional commentary tracks alongside collages of on- and off-screen voices were more frequent than synchronously recorded speech. In documentary film, speaking subjects emerge much later than in features or newsreels. Indeed, there was profound ambivalence toward giving individual embodied speech a central place. Instead, the suggestive sounds and noises of a world were intertwined with scripted poetic and collective voices. Individual embodied voices do appear, but hesitantly, with a mix of working-class subjects always discursively framed by culturally sanctioned public servants.

Keywords: non-fiction, embodied speech, subjectivity, testimony, historical reception

It is at the moment when the sync-sound embodied speech[1] emerges as a technological and ideological possibility that its stakes, limitations, and possible futures become visible. While feature films and newsreels almost immediately exploited the use of sync-sound technology, filmmakers working in the field of "documentary" as first used in English by John Grierson to refer to a kind of non-fiction filmmaking – focused their creative

1 I am using this inelegant expression to refer to film's new possibility of representing speech through synchronized images and sounds of humans speaking on-screen, whether directly to the camera/audience or to each other.

Wiegand, D. (ed.), *Aesthetics of Early Sound Film: Media Change around 1930.*
Taylor & Francis Group, 2023
DOI 10.5117/9789463727372_CH10

energies on other aspects of cinematic soundscapes. Orchestrated sounds and noises, music, and commentary tracks alongside collages of on- and off-screen voices were more frequent than synchronously recorded speech. In Anglophone documentary film, speaking subjects emerge later than in features or newsreels. Indeed, there was profound ambivalence toward giving individual embodied speech a central place in early documentary of the 1930s. Instead, the suggestive sounds and noises of a world were intertwined with omniscient and collective voices, expressed through scripted and performed polyvocal and poetic soundtracks. Individual embodied voices do appear, but hesitantly, and with a mix of working-class subjects always discursively framed by culturally sanctioned public servants.

This chapter is part of a longer examination of the genealogy of the 'talking head' and embodied testimony in nonfiction cinema. This incorporation of voices takes on varied forms, political aspirations, and meanings with different historical moments, social crises, and types of media (print, radio, film, digital video, and now even in VR). Here I explore the earliest uses of embodied speech in newsreels, news magazines, and documentary film. In the late 1920s and early 1930s, the new possibility of sync-sound vocal address to a distant audience was a practice with affective and epistemological effects that would be put to multiple uses. The reasons for the documentary's hesitant use of sync-sound in the early and mid-1930s are many. I look at the tendentious use of speech and interviews in early newsreels followed by the ambivalent and contrasting attitudes of early documentary filmmakers to sound and filmed speech.

Before Warner Brothers released *The Jazz Singer*, the first feature film containing scenes with sync-sound dialogue, in 1927, Fox Movietone had produced its earliest sound newsreels. Among Movietone's first productions, released in 1927 and 1928, were the public speeches of President Coolidge and Charles Lindbergh upon Lindbergh's return to Washington after his transatlantic flight; Benedetto Mussolini addressing the American people;[2] and playwright George Bernard Shaw mocking both Mussolini and the talking newsreel itself in a delightful performance explicitly directed to a cinema audience. Introduced by a title card as "the world's outstanding literary genius," Shaw is first seen walking down a garden path, ostensibly oblivious to the newsreel apparatus, when, as if suddenly surprised on a walk, he begins to address us. "Have you all come to see *me*, ladies and

2 The first example of direct address solely to a camera was Mussolini's brief speech, read from title cards, shot in May 1927 (Fielding, *American Newsreel*, 164), and first projected before Murnau's *Sunrise* in September 1927 (Crafton, *The Talkies*, 94).

gentlemen? Well, I should never have expected this," he says, moving his eyes back and forth across the imagined cinema crowd. Continuing to hold forth on this strange encounter, he reflects on posing for a camera and recounts an anecdote about a child autograph-seeker. Shaw's address reflexively draws attention to the limits and potential of the new medium. Mimicking Mussolini, with his "imposing look that terrifies one," as Shaw puts it, the Irish writer pokes fun at the Italian leader. Like Mussolini, Shaw began by approaching the camera; but rather than pretending it was not there, he acted as if a huge audience had shown up in his garden. At the end of the reel, about to take his leave, he says good night, but then corrects himself: "By the way, this may be a matinee, good-night may not be the right thing to say. However, call it 'good afternoon,' 'good day,' and anyhow, good-bye and good luck."[3] In a second newsreel two years later, Shaw, introduced as the "most brilliant mind in the world today," again directly addresses the movie audience by poking fun at the particularities of sound film: "Here am I, enjoying myself, down at Malvern and there are you, enjoying yourself by allowing a photograph to talk to you. That, you see, is one of the marvels of the Movietone. It brings us together in a way that used to be entirely impossible."

Indeed, the talking newsreel must have been astonishing, and early attitudes to this possibility of simultaneously hearing and seeing public figures up close were enthusiastic. The poet and film critic HD, who was extremely damning in her opinion of the use of sync-sound dialogue in dramatic films, wrote glowingly about being able to hear and see US President Coolidge and Charles Lindbergh in Fox's newsreel. For HD, having access to sync-sound voice opened a new door to comprehending the world. She writes:

> If [Movietone] were used properly there would be no more misunderstandings. [...] Nations are in turns of wrists, in intonations of voices, and that is where the Movietone can do elaborate and intimate propaganda. Peace and love and understanding and education could be immensely aided by it.[4]

3 Shaw insisted on 'directing' the newsreel himself. Though audiences and the press praised his words and wanted more, "interviews of a longer and more substantial kind did not follow. Instead newsreel producers were satisfied to provide what was really only a continuation of the same sort of novelty celebrity footage that had filled the silent newsreels of previous years." (Fielding, *American Newsreel*, 166–67).

4 HD, "The Mask and the Movietone," 209.

The talking newsreel was transformative for her, creating a radical shift in the possibilities of global understanding. It is notable that in her description of speech, she includes the audible qualities of voice, accent, and intonation and also the visual representation of gesture. Indeed, these qualities are emphasized far more than the actual words themselves, which she barely mentions.

Given the popularity of the first sound newsreels, Fox Movietone immediately recorded a number of important personalities speaking to the camera, among them Maréchal Foch, King George V, and John D. Rockefeller. Less about promoting "peace and love" than getting people into theatres, public figures addressing the audience quickly became a regular feature of newsreels. By 1929, Fox was putting out four all-sound Movietone newsreel editions each week and soon most of the other studios were also producing talking newsreels. The *New York Times*' Movietone reviews of 1930 and 1931 mention, among others, "a pictorial and sound interview with Mother Jones" on her 100th birthday, with Ethel Barrymore on the condition of theatres, and with Sinclair Lewis on winning the Nobel Prize. Mussolini was shown and heard declaring that Italy would never start a war, and the Prince of Wales spoke in "the first intimate interview he has granted the sound newsreels." Hearst newsreels included interviews with Mahatma Gandhi, Leon Trotsky, H. G. Wells, and others of more prurient interest, such as the girlfriend of recently assassinated gangster Legs Diamond. Figures were solicited for opinions or statements about the state of the world, others simply out of public curiosity about their private lives, and all for the excitement of hearing and seeing a public figure speak. Publicity frequently emphasized what was sensational: Gandhi's is "the voice of the most remarkable man in the world today," Rockefeller's is the "first appearance in sound films of world's richest man," and Rabindranath Tagore is "another of the world's outstanding figures." According to Donald Crafton, within months of their debut, many of the Movietone camera operators "became precursors of paparazzi," and the content of talking newsreels rapidly shifted "to what we call now 'soft' or even 'tabloid' news."[5] By 1932 much of newsreel sound was studio-recorded, with bogus sound effects, and occasional "fraud and faking."[6]

Following the interest in the powerful, sensational and sometimes inspirational figures of the day, the turn to person-on-the-street interviews first appeared on the radio in 1932, before the Roosevelt versus Hoover

5 Crafton, *The Talkies*, 96, 99.

6 Fielding, *American Newsreel*, 168.

presidential election in the United States. The successful broadcast subsequently became the "Vox Pop" radio program, a popular weekly show.[7] In newsreels, this approach of getting 'regular' people's opinions on the screen had its notorious first appearance in MGM's series of 1934 newsreels called *California Election News*. MGM's "Inquiring Cameraman" supposedly travelled the state getting opinions on the upcoming gubernatorial election of 1934. Yet the interviews were actually scripted and acted, to create what historian Arthur M. Schlesinger Jr. called "the first all-out public relations Blitzkrieg in American politics," Hollywood's concerted attack on leftist gubernatorial candidate Upton Sinclair.[8] In a deft piece of negative propaganda, a supposed cross-section of California – manual workers, unemployed migrants, housewives, businessmen, lawyers, and immigrants – are asked who they are voting for. Their 'spontaneous' replies, performed as convincing depictions of natural speech, subtly suggest the dangers of a left-wing Sinclair victory: less reputable and some explicitly "communist" characters saying they will vote for Sinclair while the most middle- and professional-class citizens say they will vote for conservative Merriam. (Sinclair lost.) Other newsreels would follow suit in creating cleverly edited pieces of supposed public opinion.[9] The 'candid' speaking subject-on-the-street was useful as a device to *manipulate* opinions rather than to investigate them. The scripted performance of ordinary speech had become an effective tool of political propaganda.

Like the MGM newsreels, the *March of Time* news magazine also relied exclusively on performed and scripted speech.[10] Developed from the *March of Time* radio broadcasts in 1935, it reached over twenty-two million viewers by the end of 1937.[11] The radio broadcast, originally conceived as an advertisement for the printed *Time Magazine*, employed professional actors to perform sketches illustrating news items. The actors were trained "to

7 Newspapers had frequently conducted opinion polling before elections, but this is the first known radio polling. A few years later "Vox Pop" became a quiz show sponsored by corporations, with interviewees pre-selected and prepared, rather than spontaneously found on the street.

8 Mitchell, *Campaign of the Century*, 369. See also Fielding, *American Newsreel*, 268–69.

9 For example, Pathé released a newsreel in which 'random' people are asked whether or not they think the United States should go to war in 1939. In a cleverly edited sequence, individual responses are reduced to shorter and shorter sound bites that end with a resounding and clearly directive "no." The reel is included in *Prelude to War* (USA 1942, dir. Frank Capra/Anatole Litvak).

10 For Raymond Fielding it was the *March of Time* that "successfully introduced and established the documentary format for film audiences in the United States" (*American Newsreel*, 73). However, the filmmakers working in the early documentary traditions discussed here actively differentiated themselves from all things "newsreel."

11 Fielding, *March of Time*, 185.

mimic exactly the voice patterns, inflections, and characteristics of the figures impersonated."[12] In other words, *March of Time* radio never solicited the speech of well-known or political figures; they simply recreated and performed it. This was considered so normal that no one except for President Franklin Roosevelt, when he received complaints about what he had 'said' on the program, is known to have objected.[13] The filmed *March of Time* did the same, combining actual footage of public figures with staged scenes in which they spoke. Dramatically presenting a handful of news items in twenty minutes every week, producer Louis de Rochement adopted a cinematic story-telling format mixing re-enactments and impersonations of famous people – including Hitler, Huey P. Long, and many others – with footage of the actual figures. Although actor-doubles were regularly employed, some subjects agreed to re-enact, or simply act, scenes written by the producers. Newsworthy scenes were performed in sync-sound as if they were real, with the doubles responding to the questions of an acted 'secretary,' 'reporter' or 'crowd.' This strategy permitted the careful control that gave the *March of Time* its crafted, dramatized flow. The invisible, omniscient "Voice of Time" (as stentorian narrator Cornelius Westbrook Van Voorhis was known) could then interpret this orchestrated world, a world in which no speech was unscripted.

Perhaps it was partly because the newsreels so quickly took up, and then so egregiously manipulated, sync-sound speech that many documentary filmmakers eschewed it, opting for more complex soundscapes in their work. For the documentary filmmakers of the 1930s, the complications of shooting sync outside a studio certainly contributed to this situation. However, the newly named documentary's aims and aesthetic principles were even more significant. For John Grierson, documentary's first well-known spokesman and the leader of the British documentary movement of the 1930s, documentary had to distinguish itself from the newsreel ("just a snip-snap of some utterly unimportant ceremony" and "purely journalistic"), and also from the "lecture film." It was defined as not "the plain (or fancy) descriptions of natural material" but rather the "arrangement, rearrangements, and creative shapings" of such material.[14] However, the filmmakers who Grierson hired and trained, and often argued with, developed varied approaches to what

12 Ibid., 14.

13 Ibid., 15. According to Fielding, *March of Time* stopped impersonating Roosevelt at all after several complaints from the White House in 1937. They were afraid that others might begin to object as well.

14 Grierson, "First Principles of Documentary," 145–46.

he called the "creative treatment of actuality." The "actuality" that was to be treated and how it was to be 'shaped' were circumscribed by Grierson's philosophy of film as social propaganda and civic education, as well as the government and commercial sponsorships (Empire Marketing Board, General Post Office, British Commercial Gas Association, etc.) that provided funding. As filmmaker Basil Wright put it somewhat equivocally, "Nobody's ever defined the word 'documentary' satisfactorily because it's a certain approach to the use of the cinema on [...] the social side of the community."[15]

Even with the shared focus on creative shapings of the "social," different ways of working developed among the group's successful practitioners. In a 1935 review Grierson implied that there were two types of documentarians working at the time: the "impressionists" and the "informationists."[16] While Grierson doesn't explain exactly what he means, the "impressionists" seem to be those who emphasized visual drama and used creative camerawork to suggest embodied experience in their films, while the "informationists" were more interested in providing what Grierson called "analysis" and presenting problem-solution frameworks. Several of the British filmmakers moved between approaches over time. The distinction between what Grierson here calls "impressionists" and "informationists" is, I argue, expressed largely through the rhetorical structures and soundscapes of the different films, specifically how they regard and deploy individual and embodied speech.

If distinguishing documentary from newsreels and journalism was central to Grierson, the use of creative sound seems to have played an equally important role for Brazilian-born Alberto Cavalcanti, who arrived on the Grierson team the same year as a British Visatone sound system, in 1934. Cavalcanti had an earlier career as a filmmaker in France, but came to work with Grierson because he wanted "to experiment in sound."[17] According to Wright, Edgar Anstey and others, Cavalcanti's knowledge, professionalism, and imagination had a profound effect on many of the group's films.[18] He is regularly given credit for several innovations, beginning with his contribution to Grierson's *Granton Trawler* (1935), made up of simple recorded and reconstructed noises of the ship and sea and fragments of conversation

15 Sussex, *The Rise and Fall of the British Documentary*, 54.

16 Grierson, "Two Paths to Poetry," 194–96. Others also use the term "impressionists" for many of the British documentarians including Paul Rotha, who was happy to have his film *Shipyard* (1935) so labelled by Grierson.

17 Sussex, *The Rise and Fall of the British Documentary*, 75.

18 See Aitken, *Alberto Cavalcanti*, 50–52, 75–82, Cox, "'There Must Be a Poetry of Sound That None of Us Knows,'" 176–77, and Sussex, *The Rise and Fall of the British Documentary*, 49–51 regarding Cavalcanti's crucial and generative role on the sound design for all these films.

between sailors with no explanatory commentary. Cavalcanti's well-known essay written a few years later, "Sound in Films" (1939), articulates his thoughts on the subject. Here he divides film sound into speech, music, and noise, noting that with the advent of sound technology "films went speech-mad."[19] While images function as statements for him, sound embodies the realm of suggestion and speaks directly to the emotions.[20] Of the three elements of sound, for Cavalcanti noise – not speech or music – is the most crucial. And the key to using and organizing noises and all sounds was to use them "nonsync."[21] Such ideas were not foreign to other filmmakers in the group and elsewhere.[22] Basil Wright had also co-written a manifesto on sound in 1934, the same year he was working on *Song of Ceylon* with Walter Leigh, a student of Hindemith, and with Cavalcanti. There he provocatively asserted that "talkies" were not really films, cinematically speaking; and that "the human voice is no greater in value than any other sound."[23] This skepticism about the value of synchronized speech in cinema – embodied in Rudolf Arnheim's declaration "dialogue paralyzes visual action"[24] – can be seen in Grierson's impressionists' giving priority to visual and aural world-building and refusing their subservience to oral explanations, statements, or dialogue.

Films to which Cavalcanti contributed significantly, such as Grierson's *Granton Trawler* (1934), Wright's *Song of Ceylon* (1934),[25] his own *Coal Face* (1936) and Watt and Wright's *Night Mail* (1936) have some of the most innovative sound designs of films of the day.[26] They all incorporated spoken word, but it was not primarily expository. There was poetry – the modernist poems of W. H. Auden sung or declaimed in *Coal Face* and *Night Mail*. There was a seventeenth-century text used as historical counterpoint for narration in *Song of Ceylon*. There were collages of vocal fragments to create textures in the sonic collage of telegraphic communications in *Song of Ceylon*, the recitation of the accidents and deaths in *Coal Face* or a list of possible letters in *Night Mail*. And finally there were tiny fragments of non-sync speech

19 Cavalcanti, "Sound in Films," 101.

20 Ibid., 109.

21 Ibid., 110.

22 Writings wary of the impact of sync-sound dialogue on the visual art of cinema by filmmakers and critics in the late 1920s and 1930s were common.

23 Wright and Braun, "Manifesto: Dialogue on Sound," 96–97.

24 Arnheim, "A New Laocoön," 114.

25 With regard to *Song of Ceylon*, Jamie Sexton argues for the primary role of Leigh.

26 Grierson in Sussex, *The Rise and Fall of the British Documentary*, 207; Cox, "'There Must Be a Poetry of Sound That None of Us Knows,'" and Sexton, "The Audio-Visual Rhythms of Modernity" each argue for British documentary sound as an originator of what would later be called *musique concrète*.

of characters on screen, but these emphasized texture and vocal quality (accents, intonations, contrast of human voice with ambient sounds) rather than having any expository function.[27] All the above voices, mostly off- but some rare on-screen fragments, were embedded in complex edited collages of the 'noises' (often improvised by the filmmakers in the studio)[28] of the worlds being depicted, or in modernist musical scores. These were suggestive rather than mimetic – both composer Walter Leigh and Cavalcanti played a role here. In addition, Cavalcanti recruited Benjamin Britten to create the complex modernist compositions combining instrumental performance with collages of simulated and actual industrial sounds in both *Coal Face* and *Night Mail*. This was the "orchestrated abstract sound that is the true complement to film" described by Wright and Braun in their manifesto.[29]

A stark contrast to this approach to documentary sound emerged in the films of Arthur Elton and Edgar Anstey in 1935–1936. Both had worked under Grierson and, like the other filmmakers, were concerned with providing visual representations of British public institutions, labour, and the working class in general. However, the *actual* voices of the working classes had been largely absent until Anstey and Elton began to include them in their films. While the other films used invisible or choral voices for narration, often written by poets, and their subjects' voices in fragmented phrases primarily for aural texture, Anstey and Elton incorporated sync-sound voices of public employees as well as poor and unemployed subjects in ways that gave them expository agency for the first time. Their films of this period – for instance *Workers and Jobs* (1935), *Dinner Hour* (1935), *Housing Problems* (1935), *On the Way to Work* (1936), and *Enough to Eat?* (1936) – experiment with different modes of engaging and deploying embodied voices. All include a clear tri-partite hierarchy of narration and narrators, with an invisible voice-of-God (or an embodied intellectual in the case of *Enough to Eat?*) that provides context and information to the viewer and

27 For example, the non-sync, over-dubbed voices of the sailors in *Granton Trawler* speak about football; the miners eating in *Coal Face* about a fire. These fragments of conversation suggest character and social background to those we see, rather than providing substance on the topic of the film. Grierson suggested that eavesdropping on conversations was "one of the pillars of our [sound] art" in "The GPO Gets Sound," 217.

28 Delightful descriptions of sound production – using typewriters, empty bottles, film projectors, sandpaper, a trumpet, and various percussive objects – can be found in Grierson's "Sound Comes to GPO," 216, and Cox, "'There Must Be a Poetry of Sound That None of Us Knows,'" 177–78.

29 Wright and Braun, "Manifesto: Dialogue on Sound," 97.

orchestrates the voices of the public officials and then the "people really concerned."[30] In so doing, they did what no other nonfiction films had done: while maintaining a clear rhetorical hierarchy, they nevertheless distribute the authority and agency to address a cinema public across diverse classes and voices. They include, to varying degrees, the voices of those whose *lived experience*, rather than *expertise*, is relevant to the problem being depicted.

Just as films construct viewing positions for their audiences they also construct *listening* positions, most obviously in documentary where we are directly interpellated by the film-text. The selection of forms and sources of address deployed in a nonfiction work are a crucial aspect of this process. If one kind of viewing/listening position is constructed by an unchallenged omniscient narrator and another by a poetic but romanticized voice speaking in the name of a community or collective, a quite different one is constructed when we hear embodied, living subjects speak directly to us. While narrators or poetic collective voices seem to ask us to submit to their omniscient authority or lyrical vision like students listening to a teacher or enthusiasts to an orator, embodied speakers, whom we see and hear, can engage us physically, affectively, as vulnerable fellow human beings in a shared world. In the British films by Anstey and Elton discussed here, constructed collective voices have been replaced by a panoply of individual voices, each with its own idiosyncrasies and areas of expertise, often directly related to the bodies we see on screen. These secondary voices represent local rather than universal knowledge, refer to specific events, speak from a first-person perspective and give a different inflection to what is being shown than what an omniscient narrator or observational footage can convey. Our perspective shifts with the shift in voice, as we engage sensuously with accents and sonorous qualities, as well as with the details and anecdotes that only a first-person perspective can provide. The world we see on screen is now inflected by the individual lived reality and vocal accounts.

Strategies for experimenting with voice were varied in these films. All five focus on an issue of contemporary society and explain either how things work, or how and why they do *not* work, but will work better in the future. Thus, *Dinner Hour* explains the importance of gas and its distribution in the daily life of urban residents; *Workers and Jobs* and *On the Way to Work* suggests possible solutions to the problem of massive unemployment

30 This is how the omniscient, invisible narrator of *Housing Problems* refers to the slum inhabitants.

in Britain; *Housing Problems* shows the deplorable living conditions in the slums and then explains how they are being improved; and *Enough to Eat?* provides shocking information on malnutrition in England and how it could be eradicated. In terms of recording and deploying multiple voices, some of the films included acted or re-enacted scenes of people (gas, bakery and other employees, unemployed men) speaking with each other in sync-sound to illustrate a problem or situation. Some included explanations and humorous deictic references by those present on the screen, but with their voice recorded afterwards in the studio – for example, comments on what we are seeing, who someone is, or how much money was paid and what someone did with it. *Enough to Eat?* was the first film to completely eliminate the omniscient, invisible narrator and instead have a well-known public intellectual directly addressing us on screen. But it was *Housing Problems*, still the best known of the group, that was the only film to have relatively lengthy sync-sound scenes in which individuals who are not officials speak directly to the camera. It was this ability to see and hear 'ordinary' working-class subjects that became the most noted aspect of all these works.

A good third of *Housing Problems* consists in sync-sound testimony by residents of East London slums, whose direct address to the camera was shot and recorded in their cramped homes. The omniscient narrator of *Housing Problems* introduces the piece with "A great deal these days is written about the slums. This film is going to introduce you to some of the people really concerned." He then introduces a local councillor whose expository voice is woven throughout the piece, directly commenting on problems and solutions over illustrative shots of the slum area. It is the omniscient narrator who introduces the tenants with phrases such as "And now for the people who live in the slums" and "This is what Mrs. Hill has to say." Four inhabitants speak directly to the camera, recounting nightmarish anecdotes about rats, vermin, and babies being bitten, while the film occasionally cuts away to illustrations of their words. After the 'solution' – new housing developments with windows that open, gas heating, and hot water – is presented by the councillor, more sync-sound testimonies by residents confirm the merit of that solution, followed by a collage of non-sync voices reiterating the horrid living condition over shots of a crowded alley-way.

Although opinions regarding the merits and problems of Anstey's approach were mixed, the foregrounding of subjects speaking about their own experiences was striking. In his favourable 1936 review, Graham Greene contrasted the film with BBC radio documentaries full of "ironed-out personalities with censored scripts."[31] For the most part the speakers look directly at the camera in medium shots with a few medium close-ups and speak in two to three long takes. Perhaps somewhat nervous, their speech nevertheless seems quite natural – possibly practiced (Anstey denied full rehearsal),[32] but clearly not scripted. Contemporaries criticized a lack of any aesthetic in Anstey's direction, but he claimed that the film's sparseness was deliberate. Interviewed decades later about the production, he noted:

> We narrowed ourselves down in *Housing Problems* to a very, very simple technique, which was open to us, at that time nobody had done it, and we gave the slum dwellers a chance to make *their own film*. [...] [W]hat we felt was 'this is *their film* not ours. We don't want any directorial intervention. Their story is strong enough by itself.'[33]

Speaking about the framing, he says: "[W]e felt that the camera must remain sort of four feet above the ground and dead on, because it *wasn't our film*."[34] Here it is not the naiveté or possible disingenuity of phrases such as "their own film" which interests me, but rather the possibility that this is how the film was experienced at the time, by Anstey, Elton and their contemporaries. In Anstey's and others' accounts, credit is always given to Ruby Grierson (John Grierson's sister, who died during the war) for her crucial role in putting the film's subjects, especially the women, at ease in front of the camera.[35] Perhaps her intermediary role, as well as the filmmakers' lack of a script, contributed to the sense that the film didn't belong to its directors.

Filmmaker and critic Paul Rotha initially took the film to task for its lack of aesthetic and dramatic power. He criticized it as a "piece of journalistic reporting" that showed that British documentarians could not create "actor[s] out of natural material,"[36] and elsewhere added that Anstey and Elton's method "deprived the documentary film of much of its cinematic quality. It became an illustrated lecture studded with personal interviews which provided 'documentary' evidence that the unseen commentator was

31 Quoted in Corner, *The Art of Record*, 66.

32 Ibid., 68.

33 Ibid., emphases mine. Anstey's comments are originally from an interview with Roger Graef for *Arena*, BBC2, March 9, 1982.

34 Sussex, *The Rise and Fall of the British Documentary*, 62.

35 Rotha, *Documentary Film*, 255; Sussex, *The Rise and Fall of the British Documentary*, 64.

36 Rotha, *Documentary Film*, 131, 182. In *Documentary Diary* (157) he noted: "Nobody pretended that this was good film-making; it was factual film reporting of a kind not before done."

speaking the truth."[37] Yet he credits the filmmakers as having "let people tell their own story without prompting or rehearsal. [...] Audiences were deeply moved by the film. Its grim authenticity shocked them. It brought them, perhaps for the first time in the cinema, face to face with unpleasant facts."[38] Following initial ambivalence, in 1938 Grierson wrote glowingly of Anstey and Elton's approach, writing "the greatest advance [...] came with two little films which, except among the far-seeing, went almost unnoticed. One was called *Housing Problems* and the other *Workers and Jobs*."[39] Like Rotha, he contrasted them to the romanticized approach to the 'common man' in other films. "*Housing Problems* is not so well made nor so brilliant in technical excitements, but something speaks within it that touches the conscience. [...] *Housing Problems* 'transforms' and will not let you forget."[40]

Housing Problems remained somewhat of an anomaly. While the films by Anstey and Elton on unemployment and on gas (*Dinner Hour*) had played with post-dubbed off-screen direct address of the subjects of the film in addition to synch-sound dialogue between characters, the idea of the documentary subjects being given a platform to address the world and to speak of their lived experience at any length was not taken up by other filmmakers. Anstey's next experiment with sync-sound address, *Enough to Eat?*, eliminates the invisible narrator, but it is more – rather than less – didactic, with biologist Julian Huxley (brother of Aldous) filmed sitting at a desk explaining malnutrition and its possible solutions. An advertisement for the film applauded "the little interviews with poor mothers struggling bravely to feed their families," labelling these "the most searching things in the picture."[41] Yet these scenes are miniscule in the scope of the film, and the women speak to social workers rather than to the audience. Other than Huxley, direct address comes from public officials: doctors, politicians, and a representative of the League of Nations. Other British documentary directors, even those formerly of the 'impressionist' approach, subsequently tried their hand at including more synchronized speech, usually in re-enacted or scripted scenes, while also re-instating invisible voice-of-God narrators in their social-issue films. For example, Basil Wright's *Children at School* (1937) and Donald Alexander and Paul Rotha's *Eastern Valley* (1937) both

37 Rotha, "Films of Fact," 165–66.
38 Rotha, *Documentary Film*, 256.
39 Grierson, "Battle for Authenticity," 215.
40 Ibid., 84.
41 Advertisement for *Enough to Eat?*.

experimented with sync-sound scenes, mostly acted by their subjects rather than in the form of testimony or direct address, except by occasional public officials.[42]

The fascination, ambivalence, and debate surrounding the use of sync-sound and embodied direct address in documentary film are instructive. In spite of early skepticism among documentary filmmakers, Shaw's claim that "talking photographs" bring people together in entirely new ways, HD's elation at the sudden sense of understanding imparted by the "intonations of voices," as well as Grierson's opinion that hearing the voices of the poor would "transform" us, indicate early recognition that embodied voices speaking from a screen *could* provide a profound experience. Yet the abuses of performed 'spontaneous' direct address in both credible and faked propaganda newsreels may have augmented documentary's distrust in the value of speech. In addition, Shaw's joke that it was "photographs" – notably *not* what Wright and others called *cinema* – doing the talking, was clearly salient to the documentary filmmakers, many of whom found sync-sound aesthetically and visually monotonous compared to the creatively orchestrated visuals, abstract sounds and modernist music in works by Cavalcanti, Watt and Wright. The early experiments of Anstey, Elton, and others with embodied voices created an aesthetic and philosophical rupture in approaches to the documentary screen. Visually compelling and suggestive rhythmic sound montage receded in nonfiction work as new subjectivities and speech entered the public sphere, eliciting other listening opportunities and affective engagement. Nevertheless, even though some experimentation continued both in Britain and the USA in the 1930s and 40s, unscripted embodied speech and direct address retained a minor place in documentaries. Nonfiction films truly shaped by the subjectivities and voices of on-screen speaking subjects did not proliferate until the 1960s when both sync-sound technology became more accessible and, perhaps more importantly, listening to idiosyncratic and non-hegemonic voices became a crucial part of the cultural ethos.

42 It is notable, however, that an archival print of *Eastern Valley* viewed at the BFI concludes with a powerful, unexpected scene of direct address. A village woman (a Mrs. Wetton) speaks to and faces the camera, with a story about her husband's new job delivering – and spilling – milk and the difficulties of raising her "kiddies." She concludes the film by asking us, the audience, to send money to support the new employment scheme. The entire scene with her was deleted in other prints and is not in the BFI's DVD version of the film included on its four-disc set *Land of Promise: The British Documentary Movement 1930–1950*, produced by Shona Barrett and Caroline Millar.

Bibliography

Advertisement for *Enough to Eat? The Nutrition Film* (1936). *World Film News* 1, no. 10 (January 1937).

Aitken, Ian. *Alberto Cavalcanti: Realism, Surrealism and National Cinema*. Trowbridge: Flick Books, 2000.

Arnheim, Rudolf. "A New Laocoön: Artistic Composites and the Talking Film" [1938]. In *Film Sound: Theory and Practice*, edited by Elisabeth Weis and John Belton, 112–15. New York: Columbia University Press, 1985.

Cavalcanti, Alberto. "Sound in Films" [1939]. In *Film Sound: Theory and Practice*, edited by Elisabeth Weis and John Belton, 98–111. New York: Columbia University Press, 1985.

Corner, John. *The Art of Record*. Manchester: Manchester University Press, 1996.

Cox, Geoffrey. "'There Must Be a Poetry of Sound That None of Us Knows …': Early British Documentary Film and the Prefiguring of Musique Concrète." *Organised Sound* 22, no. 2 (August 2017): 172–86.

Crafton, Donald. *The Talkies: American Cinema's Transition to Sound, 1926–1931*. Berkeley: University of California, 1997.

Fielding, Raymond. *The American Newsreel 1911–1967*. Norman: University of Oklahoma Press, 1972.

Fielding, Raymond. *The March of Time, 1935–1951*. New York: Oxford University Press, 1977.

Grierson, John. "Battle for Authenticity" [1939]. In *Grierson on Documentary*, edited by Forsyth Hardy, 215–17. London: Faber & Faber, 1979.

Grierson, John. "First Principles of Documentary," [1933–1934]. In *Grierson on Documentary*, edited by Forsyth Hardy, 145–56. London: Faber & Faber, 1979.

Grierson, John. "The GPO Gets Sound." *Cinema Quarterly* 2, no. 4 (1934): 215–21.

Grierson, John. "Two Paths to Poetry." *Cinema Quarterly* 3, no. 4 (1935): 194–96.

HD. "The Mask and the Movietone" [1927]. In *Red Velvet Seat: Women's Writings on the First Fifty Years of Cinema*, edited by Antonia Lant, 205–12. New York: Verso, 2006.

Leigh, Walter. "The Musician and the Film." *Cinema Quarterly* 3, no. 2 (1935): 70–74.

Mitchell, Greg. *The Campaign of the Century: Upton Sinclair's Race for Governor of California and the Birth of Media Politics*. New York: Random House, 1992.

Rotha, Paul. *Documentary Diary: An Informal History of the British Documentary Film*. New York: Hill and Wang, 1973.

Rotha, Paul. *Documentary Film*. New York: W. W. Norton & Company, Inc., 1939.

Rotha, Paul. "Films of Fact and Fiction" [1938]. In *A Paul Rotha Reader*, edited by Duncan Petrie and Robert Kruger, 161–68. Exeter: University of Exeter, 1999.

Sexton, Jamie. "The Audio-Visual Rhythms of Modernity: *Song of Ceylon*, Sound and Documentary Filmmaking." *Scope: An Online Journal of Film and Television Studies* 6, no. 2 (May 2004). Accessed February 2, 2023. https://www.nottingham.ac.uk/scope/documents/2004/may-2004/sexton.pdf.

Sussex, Elisabeth. *The Rise and Fall of the British Documentary: The Story of the Film Movement Founded by John Grierson*. Berkeley: University of California Press, 1976.

Wright, Basil and B. Vivian Braun. "Manifesto: Dialogue on Sound" [1934]. In *Film Sound: Theory and Practice*, edited by Elisabeth Weis and John Belton, 96–97. New York: Columbia University Press, 1985.

About the Author

Irina Leimbacher is Visiting Faculty at California Institute of the Arts and Associate Professor Emeritus of Film Studies at Keene State College. Her essays have been included in books on the San Francisco Bay Area Avant-Garde and filmmaker Robert Gardner. She has published articles in *Discourse, Film Comment, La Critica Sociologica, Wide Angle, Bright Lights, Camerawork* and *Release Print* and presented numerous papers at SCMS and Visible Evidence. As a curator, she has screened programmes at the Flaherty Seminar, San Francisco Cinematheque, Pacific Film Archive, UCLA Film Archive, MoMA New York, the Academy of Media Arts, Cologne, and many other venues.

11. *Die Nacht gehört uns/La nuit est à nous* and Multilingual Reception in Switzerland

Jessica Berry

Abstract: This chapter examines how two versions of one film – the German version, *Die Nacht gehört uns* (1929, dir. Carl Froelich), and the French version, *La nuit est à nous* (1929, dir. Carl Froelich, Henry Roussell) – were shown and received in Switzerland. Multilanguage versions (also known as multiple language versions or MLVs) were mostly produced to allow film production companies to market their films in different countries. In this context, Switzerland, with its multilingual tradition, is of particular interest. Here, both the German and French versions were released, according to the respective language region. The chapter shows that the MLV production wasn't just a successful practice with regards to countries of different languages, but also for multilingual countries such as Switzerland.

Keywords: multilanguage versions (multiple versions), speech in early sound film, noises in early sound film, historical reception, film theory

In Switzerland, the changeover from silent to sound film started in the summer of 1929. The Alhambra, a cinema in the French speaking town of Geneva, was the first to be refitted as a sound film theatre and made its debut in this capacity on August 2, 1929, marking the beginning of the sound film era in Switzerland.[1] Alongside shorter musical recordings and a sketch, the main film to be shown was *The Jazz Singer* (1927, dir. Alan Crosland).[2] The Capitol cinemas in the German speaking towns of Basel and Zurich

1 Hd. [= Jean Hennard], "Les débuts du film sonore en Suisse," 4.
2 Ibid.

Wiegand, D. (ed.), *Aesthetics of Early Sound Film: Media Change around 1930*.
Taylor & Francis Group, 2023
DOI 10.5117/9789463727372_CH11

followed with sound film premieres on August 13 and 14, 1929 respectively, with *Show Boat* (dir. Harry A. Pollard) as the debut sound film.[3]

One of the main challenges the film industry faced with this new medium was the distribution and exhibition of films in one language in countries of other languages. Different methods were explored in the early years of sound films to market them abroad. Other than the now well-established practices of dubbing and subtitles, production companies also adopted a method which, in film historiography, is referred to as multilanguage version (MLV) production.[4] In these cases, a film was shot in different languages, either with native speakers for each version or with multilingual actors starring in the various versions. These versions were mostly filmed scene-by-scene in all the languages on the same set. The films could then be marketed for different countries.

These developments are particularly interesting for Switzerland as a multilingual country as they meant that more than one version could be shown, unlike in most other countries. The case study in this chapter shows how this practice was implemented in the Swiss context and looks at the press reaction to the film versions discussed. The Italian speaking region of Switzerland is not addressed here, as no Italian version of the example chosen was produced. However, Italian MLV productions *were* shown in the country, such as *La canzone dell'amore* (*Love Song*) (1930, dir. Gennaro Righelli) and its German and French versions – *Liebeslied* (1931, dir. Constantin J. David) and *La dernière berceuse* (1931, dir. Gennaro Righelli).

Multilingual Film Exhibition in Switzerland

The example chosen is a multilanguage film in its German version, *Die Nacht gehört uns* (1929, dir. Carl Froelich) and its French version *La nuit est à nous* (1929, dir. Carl Froelich, Henry Roussell). The chapter looks at how it was shown in Switzerland, with emphasis on Zurich and Geneva, and how the two versions were received by Swiss film critics.[5] To this end, French and German speaking reviews of the films were researched in various regional daily newspapers. No French speaking reviews of the German language version or vice-versa, nor a comparison of the two, could

3 Ibid.

4 See the article by Maria Adorno in this volume, especially for a discussion of this term.

5 On the reception of *Die Nacht gehört uns* in Germany, see Wiegand, "Entdeckungsfahrt in die Welt der Geräusche."

be found. Consequently, whenever a French speaking article is referred to, the version being discussed is *La nuit est à nous*, whereas the German speaking reviews examine *Die Nacht gehört uns*.

The plot of the films centres on the race car driver Bettina/Bettine, who has an accident during a practice drive in Sicily and is saved by a stranger. A dramatic romance with a happy ending ensues. In the German version the protagonists are played by Charlotte Ander and Hans Albers, while Marie Bell and Jean Murat play the lead roles in the French film.

The two versions were filmed simultaneously as the Geneva film critic Jeanne Clouzot points out in her review of *La nuit est à nous*: "This film [...] was shot in Berlin and directed by Henry Roussell and Karl Frohlich [sic]. Two different versions of it exist so it can be released in either German or French speaking countries. Two casts took turns filming the same scenes on the same sets."[6]

Die Nacht gehört uns was first shown in Switzerland in Zurich on February 5, 1930, at the Scala cinema, marking the cinema's debut as a sound film theatre.[7] The film was a great success in Zurich, playing at this cinema for eight consecutive weeks, as well as being shown again at a different cinema, the Seefeld, for another two weeks in the same year. *Die Nacht gehört uns* could be seen at least once more in Zurich a few years later, in 1933, at the Olympia cinema. *La nuit est à nous* premiered in Geneva on April 3, 1930, at the Alhambra where it played for two weeks, and for another week in July of the same year.

Interestingly, the German version, *Die Nacht gehört uns*, was also released in Geneva, at the Palace cinema, where it first ran for two days in February 1931 and again for a whole week in June of the same year. It wasn't unusual for German versions of multilanguage productions to be shown in Geneva, other examples being *Leutnant warst du einst bei den Husaren* (1930, dir. Manfred Noa) at the Colisée cinema in March 1931, *Olympia* (1930, dir. Jacques Feyder) and *Ich geh' aus und Du bleibst da* (1931, dir. Hans Behrendt), both at the Palace cinema in September 1931. All these are examples of films of which a French version was also produced. In 1930 the German speakers in the Canton of Geneva amounted to some 14 per cent of the population and some 15 per cent in the city of Geneva, so that showing films in German made commercial sense.[8] In general, local populations in Switzerland were

6 J. Ct. [= Jeanne Clouzot], "De film en film," 7 (unless otherwise noted, all translations are my own).

7 Zy. [= Victor Zwicky], "Das lebende Bild," February 8, 1930.

8 Eidgenössisches Statistisches Amt, *Statistisches Jahrbuch der Schweiz*, 29, 39.

receptive to foreign-language films due to the multilingual background of the country, though the research for this chapter did not find evidence that *La nuit est à nous* was released in Zurich.

Die Nacht gehört uns/La nuit est à nous as a Breakthrough

The reviews of both language versions share the same ambivalence towards the films, though the dominant sentiment is that they represent a breakthrough, and that they are a new achievement in sound film technology and aesthetics.

The Zurich critic of the *Neue Zürcher Zeitung* says the following about the German version: "In the German spoken film 'Die Nacht gehört uns' the rendition of sounds and voices surpasses everything we have heard so far."[9] The author especially appreciates the sounds, highlighting the noise made by the motors at the car races, the whirring of the machine rooms, the clamorous jumble of sound at the horse race, etc.[10] According to this critic, as well as to the reviewer for the *Neue Zürcher Nachrichten*, these noises come very close to sounding "natural" or "realistic."[11]

Victor Zwicky, the film critic for the Zurich newspaper *Tages-Anzeiger*, also points out in his article that Die *Nacht gehört uns* is a big step forward compared to other sound films: "[A] work that is miles better than anything else like it."[12] He adds that Carl Froelich is the first German film director to produce a well-made and gripping sound film.[13]

In the Bernese newspaper *Der Bund*, the reviewer focuses on the subject matter, stating: "In any event, this represents *the* film of our time par excellence. It powerfully reveals the feeling of modern times for the machine which hisses, sings, roars, drones to become an animate living being."[14] The author feels that *Die Nacht gehört uns* is a quintessential reflection of the contemporary modern era, the era of the machine that, with all its sounds, is as much a protagonist in the film as the characters. This critique incorporates a philosophical aspect, whereas the other reviews stick more closely to evaluating the implementation of sound technology and its aesthetic impact on cinema. However, with its emphasis on the idea

9 Tp., "Zürcher Kinorundschau," 2.

10 Ibid.

11 Ibid.; Dr. E., "Filmchronik," 1.

12 Zy. [= Victor Zwicky], "Das lebende Bild," February 8, 1930.

13 Ibid.

14 Rz., "Berner Filmschau," 2.

of change, on the film representing something new, this article is in line with others, especially by emphasizing the noises and sounds heard, which fascinate all the reviewers.

In the French speaking reviews of *La nuit est à nous* this sense of the film's novelty is also present. In her article for the *Journal de Genève*, Jeanne Clouzot describes the film as the first 'true' French sound film to be shown in Geneva, as the few prior releases were either too theatre-like or too far removed from any known art concept.[15] *La nuit est à nous*, on the other hand, is 'real' sound cinema, and according to her, one of reasons is that it is not centred around a musical theme: "*La Nuit est à nous* is cinema, original sound cinema even, as the film wasn't designed to use a singer."[16]

Clouzot also sees immense progress in the fact that the dialogue doesn't slow down the action, an assessment the critic of the *Tribune de Lausanne* agrees with:

> 'Talkies' have been accused of slowing down the action and having too many musical interludes. There is nothing of this in *La nuit est à nous*: on the contrary, the film is filled with life, vibrant life, that is conveyed by the whirring of the motors, the hubbub of a factory and by the expression of anxiety, passion and joy that is reproduced so vividly.[17]

This assessment reflects the concern of critics that spoken dialogue would make films more static, partly because of the more cumbersome film equipment impeding movement, and partly because the focus would be on characters talking rather than on the action. Another concern was that the trend to filming on interior sets for the benefit of sound recording would further immobilize the medium.[18]

In the case of *La nuit est à nous*, it seems that critics accepted that their concerns were not justified and the *Tribune de Lausanne*'s journalist sees the hum of the motors and the babble of the factory rather more as the expression of life.[19]

Clouzot further claims that the spoken word is used in a new way in this film. One of the innovations the critic admires is the use of speech outside of conversations. One example she mentions is the announcements through

15 J. Ct. [= Jeanne Clouzot], "De film en film," 7.
16 Ibid.
17 P. V. [Pierre Vidoudez], "Devant les films," 4.
18 J. P. [= Jean Peitrequin], "Les cinémas," 5.
19 P. V. [Pierre Vidoudez], "Devant les films," 4.

the loudspeakers at the car race. Furthermore, she considers that in romantic scenes, silences, in alternation with dialogue, generate heightened emotion.[20]

For Freddy Chevalley, another Geneva film critic, the novelty mainly lies in the technical advances. He explains: "We should point out [...] that for the first time Carl Froelich [...] was able to make use of certain laws of acoustics, by perfectly adjusting the volume of the sound to correspond to the distance of its source."[21] Initially, this aspect preoccupied the critics and, as Chevalley's comment shows, they weren't always satisfied by the result.[22] In the case of *La nuit est à nous*, the Geneva reviewer's expectations regarding the volume-distance-correlation are met, and he also writes that "at last" the sounds are clear enough not to confuse the spectator.[23]

The articles cited show that the two versions were considered a new step forward in sound film production in both language regions. They all see the versions as progress, as a 'first,' as a breakthrough.

The Habit of Criticizing

However, despite their praise, it is also clear that critics had reservations about sound film in general. The expressions of approval are often accompanied by criticism, as in the review of the *Neue Zürcher Zeitung*: "At times, the dialogue, though quite clear, is sluggish and often suffers from an unnatural dynamic."[24] Here, satisfaction with the technical side – that is, the clarity of the dialogue – is offset by criticism of the aesthetic aspect: the author finds that the dialogue lacks rhythm and doesn't feel natural.

Zwicky in turn writes: "However, the key factor is that the technical challenge, the noiseless playback of language, noise and music, seems to be well worked out, not yet perfectly, there's still a long way to go to reach perfection, but much better than hitherto."[25] He is not completely satisfied with the technical aspects, stating that they need further improvement. However, this remark also shows that the critic is hopeful that perfection is attainable.

Contrary to Chevalley, Zwicky thinks that setting and sound, distance and volume do not correlate. He criticizes that all the characters speak at the

20 J. Ct. [= Jeanne Clouzot], "De film en film," 7.
21 F. Ch. [= Freddy Chevalley], "Devant le film," 6.
22 Müller, *Vom Stummfilm zum Tonfilm*, 287–90.
23 F. Ch. [= Freddy Chevalley], "Devant le film," 6.
24 Tp., "Zürcher Kinorundschau," 2.
25 Zy. [= Victor Zwicky], "Das lebende Bild," February 8, 1930.

same volume regardless of whether they are in the foreground or background and adds that conversations, supposedly being held outside, sound as if they are taking place in a marble hall.[26] These differences of opinion might be assumed to stem from the fact that the two reviewers write about the two different language versions. However, Clouzot also suggests that the correlation isn't always exact: "The voices sound good (why does only Marie Bell's voice sound as if it doesn't come from her mouth?) and their volume often coincides with the characters approaching or moving away. Often, I say, but not always."[27] It is therefore possible that both Clouzot and Zwicky placed more importance on the technical side than Chevalley and, consequently, were more critical. In a more general context, it can be seen that the critics of the time valued different aspects of the new advances in the art of film.

The crucial point, though, seems to be the following: "Nevertheless, I personally think that even though the scenes are skillfully done, there is too much talking in *la Nuit est à nous* [sic]. [...] One has the impression that verbal fillers, including monologues, are unnecessary on the screen."[28] Although Clouzot praises the implementation of speech, as shown above, as well as the aesthetics of the scenes, she criticizes the amount of dialogue. She, like many other critics of her time, favours a moderate use of speech in sound films.[29] Chevalley is one of these critics:

> Thus, despite all best efforts, the dialogues, and there are many, interfere with the continuity of the action. We do not contest the fact that the conversations in themselves are of interest, but they allow for so many cinematic 'malfunctions', occurring randomly, that the illusion is somewhat destroyed.[30]

While he appreciates the content of the dialogue, he feels it slows down the action and can even disrupt the illusion due to the fact that it can cause "mishaps." Chevalley is certain that audio-visual story telling will be perfected but thinks film directors and screen writers must move away from "pretentious conversations."[31] This opinion is a good example of the stance of many critics on the implementation of sound in film: they are in favour of a more creative use of sounds and speech rather than 'simple' or long-winded conversations.

26 Ibid.
27 J. Ct. [= Jeanne Clouzot], "De film en film," 7.
28 Ibid.
29 E.g., Gehri, "Où va le cinéma?," 32.
30 F. Ch. [= Freddy Chevalley], "Devant le film," 6.
31 Ibid.

All the reviews display a combination of euphoria and criticism, the latter being quite drastic in some of the articles. Some critics refer to the technical aspects, some to speech inhibiting the action. However, all reviewers feel the need to voice criticism. These conflicting views seem to be symptomatic for the time. At the time when *Die Nacht gehört uns* and *La nuit est à nous* were released to Swiss cinemas, sound film was still a novelty in the country and most critics had strong reservations about the unfamiliar medium, especially with regards to dialogue. Many film reviewers, not only in Switzerland but also internationally, felt the art of storytelling through film images had been perfected in silent films and that this cherished aesthetic no longer had a place in the new medium. They worried that in sound film the primacy of the visual would be eclipsed by the spoken word, so that it became an approximation of the theatre. This would jeopardize its special appeal and its claim to being an autonomous art form.[32] Noises on the other hand were more readily accepted because they didn't interrupt the storytelling to the same extent as dialogue and, in addition, they weren't associated as strongly with 'canned theatre.'

The reviews of *Die Nacht gehört uns* and *La nuit est à nous* clearly reflect these reservations, as well as the excitement regarding the possibilities of sound film. They agree that it is just a matter of time until sound film finds its way as a new medium, and all agree that *Die Nacht gehört uns* and *La nuit est à nous* are a big step in the right direction. The following quote from Clouzot's article is representative of this sentiment: "While the talking film has yet to find its feet, this is one of its cleverest examples so far, which inspires confidence for the future."[33]

These remarks show that film critics at the time had a particular idea of what sound film was supposed to look and sound like. Some expected the representation of noises and voices to be natural and realistic, others emphasized that the dialogue and sounds should not undermine the film image. In essence, these critics were writing their own small-scale film theories on sound film aesthetics.

The Joy of Films Speaking One's Own Language

One aspect, which interestingly is only mentioned in the French speaking reviews, pertains to the pleasure of watching films in their own language.

32 See, for instance, Rossholm, *Reproducing Languages, Translating Bodies*, 54.

33 J. Ct. [= Jeanne Clouzot], "De film en film," 7.

Jean Nicollier writes in his review for the *Gazette de Lausanne*: "[W]hat a great pleasure to hear pure spoken French after so much New York or California twang."[34] The first (talking) sound films in Switzerland, but also Europe in general, were American, and they were released in their original language. This resulted in a high demand for films in the mother tongue and a rejection of American films, as demonstrated in Nicollier's statement, in which he disparagingly characterizes the voices as "twang."

The critic of the *Feuille d'avis de Lausanne* is also extremely enthusiastic about the new possibility of seeing and hearing French speaking films:

> And, readers, do you know what causes me infinite joy about talking films? At long last, French speakers like us will be able to watch films made for us, and we will be able to hear the best Parisian French [...]. Beautiful French, nothing less, and if only for this I would welcome the new invention with open arms![35]

It seems that not only the language itself is of importance to the author, but also the cultural context of the film's genesis and its destined audience. The prospect of seeing films made with French casts for French people pleases them greatly and they predict that in future seeing foreign-language films will be a rare exception. It appears that the reviewer is not aware of *La nuit est à nous* having a German background, as they add: "[A] very fine French work, which is an inspiration for us; let us congratulate our neighbours for having achieved such an accomplished film at their first attempt: how gratifying for them and us."[36] Chevalley, on the other hand, doesn't specifically mention that the film is in French, but rather that it is of European provenance: "A European sound film which undoubtedly is far superior to talkies from the other side of the Atlantic."[37]

Conclusion

Even though critics were sceptical about the new 'sound film' medium, the release of films in their own language or even simply made in Europe greatly promoted its acceptance. *La nuit est à nous* and *Die Nacht gehört uns*

34 J. Nr. [= Jean Nicollier], "Parmi les films," 2.

35 P. [= Maurice Porta], "Dans nos cinémas," 2.

36 Ibid.

37 F. Ch. [= Freddy Chevalley], "Devant le film," 6.

played a key role at the beginning of sound film, because they were among the first films in German and French. The excitement about this fact is very clear in the reviews of both versions.

This shows that, in Switzerland, the concept of releasing individual versions for specific language regions was a success. As mentioned above, the country was in the unique situation of being able to show the various versions of one film project due to its multilingual context. This study has shown that this opportunity was seized, and in certain cases assured success. In the case of the Lausanne critic quoted, the purpose of multilanguage production – to create films that work in different national or linguistic contexts – was so well implemented that they weren't even aware of the version's German background. Clouzot, the Geneva journalist, mentions the two versions of the film – but doesn't review the German version shown in the Palace cinema nor does she compare it to the French film. The German speaking reviewers do not refer to the French version at all. This highlights the fact that, despite Switzerland's exceptional position where both films were shown in one country and in the case of Geneva even in one town, the critics didn't necessarily watch or even know about the other versions. It seems plausible that the journalists in the German speaking region were less interested in other versions of films that originated from Germany. Although the concept of an 'original' and a 'translation' is at the very least problematic in the case of multilanguage films, it is likely that some critics perceived 'their' versions as the 'originals' and consequently were not as interested in the respective 'translations.'

However, the versions of other multilanguage productions *were* compared with each other in the Swiss press, with some critics emphatically welcoming the opportunity.[38] The practice was still very new in Switzerland when *Die Nacht gehört uns* and *La nuit est à nous* were released. It is possible that critics weren't familiar enough with it and didn't yet recognize the possible worth of comparing the versions – as they subsequently did. Furthermore, the two versions were shown with a ten-month-difference in Geneva, which might have played a role.[39]

38 See B. [= William Bernard], "Les films de la semaine," 5; Zy. [= Victor Zwicky], "Das lebende Bild," October 25, 1930.

39 However, in the case of the early multilanguage production *Atlantic/Atlantik/Atlantis* (1929, dir. E. A. Dupont), a Geneva critic compares the French with the German version even though they were shown ten months apart.

This case study has shown the potential of examining the release of multilanguage productions in Swiss cinemas. Aspects touched upon here can be explored in depth when looking at further examples. In the first years after their release, *Die Nacht gehört uns* and *La nuit est à nous* were often referred to in Swiss film magazines as the first successful sound films, showing that they had a huge impact on sound film reception in Switzerland. What was said and felt about the two versions is representative of reactions to the new medium in general. The excitement and concerns manifested can be seen as an overall statement on sound film at that time. As such, *Die Nacht gehört uns* and *La nuit est à nous* are of great importance to film historiography in Switzerland.

Bibliography

B. [= William Bernard]. "Les films de la semaine." *La Tribune de Genève*, September 16, 1931.

Dr. E. "Filmchronik." *Neue Zürcher Nachrichten*, February 8, 1930.

Eidgenössisches Statistisches Amt = Bureau Fédéral de Statistique. *Statistisches Jahrbuch der Schweiz 43. Jahrgang = Annuaire Statistique de la Suisse 43me année*. Basel: E. Birkhäuser & Cie, 1935.

F. Ch. [= Freddy Chevalley]. "Devant le film." *La Suisse*, April 7, 1930.

Gehri, Alfred. "Où va le cinéma?" *Feuille d'avis de Lausanne*, September 19, 1930.

Hd. [= Jean Hennard]. "Les débuts du film sonore en Suisse." *Schweizer Cinema Suisse*, September 1, 1929.

J. Ct. [= Jeanne Clouzot]. "De film en film." *Journal de Genève*, April 6, 1930.

J. Nr. [= Jean Nicollier]. "Parmi les films." *Gazette de Lausanne*, April 7, 1930.

J. P. [= Jean Peitrequin]. "Les cinémas." *La Revue*, March 19, 1930.

Müller, Corinna. *Vom Stummfilm zum Tonfilm*. Munich: Wilhelm Fink, 2003.

P. [= Maurice Porta]. "Dans nos cinémas." *Feuille d'avis de Lausanne*, April 28, 1930.

P. V. [= Pierre Vidoudez]. "Devant les films." *Tribune de Lausanne*, April 20, 1930.

Rossholm, Anna Sofia. *Reproducing Languages, Translating Bodies*. Stockholm: Stockholm University, 2006.

rz. "Berner Filmschau." *Der Bund*, May 2, 1930.

tp. "Zürcher Kinorundschau." *Neue Zürcher Zeitung*, February 7, 1930.

Wiegand, Daniel. "'Eine Entdeckungsfahrt in die Welt der Geräusche'. Zur historischen Rezeption des frühen Tonfilms *Die Nacht gehört uns* (D 1929)." *Montage AV* 30, no. 2 (2021): 59–76.

Zy. [= Victor Zwicky]. "Das lebende Bild." *Tages-Anzeiger*, February 8, 1930.

Zy. [= Victor Zwicky]. "Das lebende Bild." *Tages-Anzeiger*, October 25, 1930.

About the Author

Jessica Berry is a PhD candidate in film studies at University of Zurich. Her thesis focuses on how the changeover from silent to sound film took shape in the multilingual space of Switzerland with the specific challenges posed by multilingualism for sound film in the period 1929 to 1934. Her master thesis on multilanguage versions in the early sound film period has been published as *Kino der Sprachversionen: Mediale Praxis und Diskurse zu Beginn des Tonfilms, 1929–1933* (2018).

12. The Mimetic Attempt of Multiple Versions

Language, Voice, and Transcultural Talkies (1929–1932)

Maria Adorno

Abstract:
As the first systematic audio-visual translation strategy in film history, multiple versions (MVs) are symptomatic of the media change of the early 1930s. In a first step, this chapter explores the historical context of MVs, focusing on the film industry's approach to the (inter)nationalism of sound films in Europe. It then investigates the aesthetic specificities of this strategy: the goal of translating between different national mentalities; the films' overcoming of the dualism between "original" and "copy"; and their role in the emergence of a "vocal aesthetics." Through a comparative analysis of trade press articles and films, the chapter argues that MVs should be understood as a "mimetic" technique that operates transculturally on multiple levels.

Keywords: multiple versions (multilanguage versions), speech in early sound film, voice, mimesis, historical reception

The acoustic materialization of languages and voices with the coming of sound made language a true challenge for the film industry, as it had to find a way to overcome national borders and linguistic diversity. As the first systematic audio-visual translation strategy and the first effective answer to the linguistic challenges linked to the dissemination of sound film, "multiple versions" (MVs)[1] are a film-historical phenomenon deserving of special attention.

1 What I call MVs was referred to in many other ways in the early 1930s, e.g., "multi-versions," "language versions," "foreign versions," "international versions," "versions-for-abroad,"

Wiegand, D. (ed.), *Aesthetics of Early Sound Film: Media Change around 1930*.
Taylor & Francis Group, 2023
DOI 10.5117/9789463727372_CH12

Although MVs have their roots in the silent era, it is with the proliferation of talkies that they irrupt into the universe of the seventh art as a systematic transnational strategy.[2] MVs involved most European countries, especially Germany and France, and diffused intercontinentally. During the peak period between 1929 and 1932, about 1,500 MVs were (co)produced internationally, involving up to fifteen languages, as well as hundreds of directors and film stars.[3] Contrary to general assumptions, MVs were not limited to musical films but rather involved all types of genres, from drama to thriller, from comedy to war or action, and even mountain films. The extent of the phenomenon testifies to how positively accepted MVs were; it also makes MVs a symptomatic, essential manifestation of the media change of the early 1930s.

With this overview in mind, I explore MVs within the European context, initially focusing on how voice and language were approached in relation to (inter)nationalism. The second part of the article highlights the specificities of what I call the "polyhedric" strategy of MVs (referring to the idea that each film shot in MVs is composed and defined by these versions, as a polyhedron is by its faces): their process of translating between different national mentalities; their intrinsic questioning of the relation between an 'original' and its 'copies'; and their role in the emergence of an acoustic/vocal aesthetics, which has been little explored so far.

To illustrate these arguments and interconnections, I investigate the trade press discourse and a variety of case studies with a comparative approach. The analysis centres on material from France and Germany, and partly also from Italy, another country that played an important role in

"export-versions" – a variety that mirrors the ambiguity surrounding the phenomenon. Modern-day scholars often call them "multiple-language versions" or "multilanguage versions" (MLVs), thereby focusing on the linguistic element. Several reasons lead me to choose otherwise: the term "multiple versions" occasionally appears in the press of the early 30s, while, to my knowledge, "multiple-language versions" does not; "multiple-language" could be mistaken to mean a film that simply includes dialogues in several languages; a 'silent' version was often planned and shot alongside the 'sound' versions; there is no good translation for "MLV" in French (nor in Italian!); and most importantly, as this article aims to stress, MVs involve a variety of cultural aspects *other* than languages. Another thing to clarify from the beginning is my approach to mentioning film titles: for reasons that are discussed in this text, I write the titles of all versions when referring to a film *tout court*, and just one title when referring to a specific version.

2 Examples include multi-camera shooting and even the versioning process itself. *La Sortie de l'Usine Lumière à Lyon* (*Employees Leaving the Lumière Factory*) (1895, dir. Louis Lumière) was already shot in three versions, though here the variations are not transculturally motivated as in the case of later MVs.

3 These are my personal calculations based on my research and on the databases created between 2004 and 2006 by CineGraph (Hamburg) and Gradisca School (Udine).

the phenomenon. Looking at the theory and practice of MVs, this article contributes to the complex mapping of their polyhedric practice – an essential step towards a systematic understanding of this translation strategy.

Sound, Language, Voice, and (Inter)nationalism

> The international and universal nature of the screen will be lost through speech.
> – Alexandre Arnoux[4]

An immediate concern arising with the dissemination of sound film is that of a potential reversion to national films. Indeed, in this phase, silent films become linked to the idea of a universal cinema, whilst sound films appear as nationally oriented. This was particularly evident within the European context, as Europe was the only continent where many countries already had prolific film industries, while also sharing borders and speaking a variety of languages, mainly on a national scale.

Thus, it is important to stress that, from 1930 on, cinema "starts to 'write' its own history" by looking back at silent cinema.[5] This ontological and historiographic crisis, to use Rick Altman's terminology,[6] emerges in the press of each country: initially, the situation is portrayed as a battle between silence and sound but also – more specifically related to the linguistic issue – as an unsolvable conflict between (American) English and all other languages. Articles against sound film proliferate, depicting sound as a threat to the 'pure universal cinematic art,' which – unlike sound film – is "graphic and visual."[7]

Thomas Elsaesser has highlighted how language is simultaneously an "un-crossable barrier and an un-missable supplement" of the cinematographic medium.[8] Indeed, a complementary reaction emerged almost immediately, aiming to foster sound films rather than oppose them, despite the linguistic barriers. There were at least two reasons for this, one aesthetical and one economical.

4 Arnoux, "Il peut se créer un art neuf." If not indicated otherwise, all translations are mine.

5 Szczepanik, "The 'Birth' of Cinema History," 33. Szczepanik's analysis focuses more specifically on the rise of the first film archives and of a historical self-consciousness of cinema as art.

6 Altman, "Penser l'histoire du cinéma autrement," 65–74.

7 Arnoux, "La bataille des langages et le film parlant."

8 Elsaesser, "ImpersoNations: National Cinema, Historical Imaginaries and New Cinema Europe."

With respect to aesthetics, it became evident that 'silent' films were actually not *that* international. A hint in this direction comes from a contemporary full-page article on a lesser-known case of MVs: *L'Étrangère/Die Fremde/La straniera* (*The Foreigner*) (1930, dir. Gaston Ravel/Fred Sauer/Amleto Palermi), based on the novel by Alexandre Dumas. The three versions were shot in Paris, as a coproduction of the German Hegewald and the French Jean de la Cour companies. Italian film diva Francesca Bertini was supposed to star in all three versions, but they ended up having different casts. In his very attentive comparative analysis of the casts' costumes, gestures, and expressions in an article in 1931, Roger Régent points out several differences: *La straniera* has the character of a tragedy, *L'Étrangère* has a lighter, joyful style, and *Die Fremde* is serious, though not tragic. These observations might be influenced by national stereotypes, though the author's arguments are supported by visual material from a romantic scene, and he offers a valid argument:

> And now, who will dare to say that the image alone was international and that the word given to cinema limits its expression? [...] The way of giving a necklace is not the same in Berlin as in Paris or Rome. Nor is the way of receiving it, and yet, in Berlin, Paris, and Rome, there is still only a woman and a necklace ...[9]

The author suggests that it is not only sound that nationalizes a film. Instead, the visual component of a film can also be culturally specific and judged according to the context where it is shown.

With regard to economics, European countries soon realized that none of their film industries would be able to survive only on the national markets. Import and export had to be promoted so as not to lose the internationalism of film. This sentiment is mirrored by the press, as publications switch their focus from a rhetoric of fear to one of hope. For instance, in an article in *La rivista cinematografica*, sound film does not appear entirely dangerous anymore, and we learn that there are three "genres" of sound films: the "film with sound effects," the "scored film," and the "talkie" (*films sonoro, musicale e parlato* in the original Italian). The only one posing problems to the film industry, according to the text, is the third one.[10]

9 Régent, "France, Allemagne, Italie."
10 "Il film internazionale finirà?"

In fact, even though many texts mention the linguistic issue, retrospectively, we can see that language was not the real problem, or at least not the main one. As Chion argues, the problem was "the voice, as material presence."[11] Language barriers had also played a role in silent cinema and needed to be addressed when exporting films – at least to a certain extent. What early sound film historiography must confront, then, is the emergence of the vocal element, in other words, the emergence of the *spoken* language. In terms of film technology, the novelty was not even the speaking, for it is now well recognized that actors also spoke on silent film sets; the real novelty was in sound recording technology and thus in *recorded* spoken language, in the ability to register the speech act and reproduce it in cinemas.

Joining the debate on the (inter)nationalism of sound films, many authors writing in the German, French, and Italian trade press start to introduce the same revolutionary idea with reference to films such as *Melodie des Herzens* (*Melody of the Heart*) (1929, dir. Hanns Schwarz), *La canzone dell'amore* (*The Song of Love*) (1930, dir. Gennaro Righelli), and, especially in France, the new productions by Joinville Studios. The idea that emerges is that of "multiple versions."

MVs in Theory and Practice: Towards Transcultural Talkies

MVs reshoot the 'same' film for each country interested in its production in a different language, based on the same screenplay, often at the same time, and often on the same set, potentially adapting all elements of the film to the respective target context, starting with the cast.

This definition sounds quite straightforward, but anybody who is familiar with the topic knows that there is an extremely large variety of cases. Among other things, not all studios shot MVs in the same way: Ufa specialized in faithful 'versioning,' that is, shooting the versions scene-by-scene, while Paramount in Paris used a film-by-film method, which was faster but less precise. Next to these major studios, both the French Vandal & Delac and the Italian Cines, for instance, produced about ten films as MVs in these years, paying as much attention to detail as Ufa.

Altogether, MVs introduce a complex, expensive procedure, as the shooting of any additional version substantially increase the amount of

11 Chion, *The Voice in Cinema*, 12.

time and finances required. Despite these obstacles, MVs were produced systematically at least until 1931 in the USA and during the first half of the 1930s in Europe. Also, it is important to stress that, contrary to common assumptions, MVs never really disappeared. Not only did they continue until the 1970s,[12] but there are contemporary cases, too.[13] This likely has to do with the advantages they offer, as thoroughly analysed by Joseph Garncarz. First, MVs demand no learning process on the part of the audience: unlike dubbing, they require no suspension of disbelief with regard to the actors' body-voice unity (an aspect Erich Pommer often hinted at);[14] and, unlike subtitling, they do not require any reading effort. Also, and most importantly, MVs answer to the needs of different populations and cultures by removing the supposedly *foreign* elements and replacing them with familiar ones. In other words:

> The problem of exporting films is connected with the different tastes of the culturally differentiated national audiences in Europe. [T]he reason why multiple-language versions had a chance of showing a profit even at a time when dubbing had come to be accepted was that they responded not only to linguistic diversity, but were also an effective response to the problem of cultural diversity.[15]

Language, cast, and crew, but also locations, gestures, dialogues, expressions, clothes, music, lyrics – all these elements could be varied depending on the version and on what was expected to be unacceptable or unintelligible for the audience of a specific version. Among all the components of a film that might connote cultural foreignness, the cast is probably the most prominent one and the most essential to adapt in order to guarantee the success of a version. Casting choices were based on the language skills of actors and actresses but also on the fact that cinemagoers wanted to see *and* hear their own stars directly. As suggested by Ivan Klimes and Pavel Zeman, language has a powerful symbolic dimension, so that "the fact *that one is communicating in a particular language* can at times be more important than *what one is communicating in that language*."[16] Film stars such as Lilian

12 Barnier, "Versions multiples et langues en Europe."

13 For instance, the English-Bosnian *In the Land of Blood and Honey/U zemlji krvi i meda* (2011, dir. Angelina Jolie) or the first and only TV series ever shot in MVs, the British-Welsh *Hinterland/Y Gwyll* (2013–2016, dir. Ed Talfan and Ed Thomas).

14 Pommer, "Pommer über den mehrsprachigen Tonfilm."

15 Garncarz, "Making Films Comprehensible and Popular Abroad," 74.

16 Klimes and Zeman, "Aktualita 1937–1938," 62. Italics in the original.

Harvey, Greta Garbo, Henry Garat, or Maurice Chevalier truly functioned as symbolic cultural vehicles.

Translating Mentalities

A love scene never has the same tonality in Berlin, Paris, or London.
– Erich Pommer[17]

Ufa's head of production Erich Pommer soon starts to clarify his position in favour of international sound films, supporting the emerging idea of the MVs and of a "transfer" of mentalities:

> The new art form – the sound film – needs a new dramaturgy. [...] A literal translation must be avoided at all costs. Instead, a transfer from one mentality to the other must be sought. [...] The different versions must not only do justice to each national language but must also correspond to the diverging mentality in the image composition and the execution of the individual scenes.[18]

Indeed, trade press articles – especially in Germany – often stress that these *multiple* productions should take into account the mentality of the various countries. For instance, managing director Ludwig Klitzsch, in line with Pommer's philosophy, emphasizes how MVs must be planned with view to the specific mentalities of the countries concerned from the very beginning, and how film is a "national cultural" vehicle.[19] But it is the French magazine *Pour Vous* that links MVs and the issue of language directly to the audience, saying that the dialogists of film versions had to take into account "the differences in the mentality of two audiences."[20] Therefore, a central aspect of the MVs phenomenon, which connects all of them despite the variety of approaches, is the intention to shape a film's versions on a cultural level, to 'multiply' the film and yet to keep it 'identical.'

Numerous cases could show how linguistic adaptation was used to shape the "mentality" behind each film produced as MVs, such as in *Der Kongreß tanzt/Le congrès s'amuse/The Congress Dances* (1931, dir. Erik Charell), the

17 Pommer, "Pommer über den mehrsprachigen Tonfilm."

18 Ibid.

19 Klitzsch, "Der Film als nationaler Kulturfaktor."

20 Natanson, "Les joies et les ennuis du dialogueur."

Figure 12.1: Postcard for *Der Kongreß tanzt/Le congrès s'amuse/The Congress Dances* (1931, dir. Erik Charell) showing the three actors for the role of Pepi/Pépi as well as Lilian Harvey, who played in all three versions.

most popular film in Germany in 1931/32.[21] The three versions appear almost identical at first glimpse,[22] which can be attributed to Lilian Harvey playing the main character in all of them and to Pommer wanting to prove "that we in Germany are capable of producing foreign-language films."[23] In this sense, the film is an excellent example of Ufa's scene-by-scene method. A film postcard by Ufa illustrates this: the three actors interpreting the friendly character of Pepi/Pépi not only wear almost identical clothes and have similar postures, they also look alike, especially Robert Arnoux and Carl-Heinz Schroth (Fig. 12.1). This postcard is also a visual confirmation of how MVs can play on multiplicity: the *one* Christel interpreted by polyglot Harvey is paired with *multiple* 'Pepis' played by their respective German, French, and English actors.

When considering the characters' identities, then, it is important to keep in mind that each of the characters at the eponymous Congress of Vienna, which the film depicts, represent a specific country or territory

21 Garncarz, *Wechselnde Vorlieben*, 208.

22 A detailed comparative analysis of these three versions can be found in Claus and Jäckel, "*Der Kongreß tanzt*: Revisited."

23 Quoted in Hardt, *From Caligari to California*, 136.

(both in fiction and reality). Nationally and culturally specific elements are not absent but rather carefully delivered, often employing the tool of irony. The film simultaneously draws on national/cultural differences *and* similarities. For example, by eliminating the 'bad' joke about French men always being late "because of women," *Le congrès s'amuse* avoids offending the French audience, whereas by keeping it, *Der Kongreß tanzt* and *The Congress Dances* give the Germanophone and Anglophone audiences an opportunity to make fun of the 'others.' Mutual substitutions are even more interesting, as in the case of a short scene at the council: in *Der Kongreß tanzt* we see and hear the King of Saxony ("König von Sachsen") commenting on how much better the coffee is "here in Dresden" ("bei uns in Dresden") compared to 'there' in Vienna, while in *Le congrès s'amuse* it is the French Minister who speaks, making a joke about the Tsar of Russia instead. Also, a wider analysis of the songs and their respective lyrics would show how *Le congrès s'amuse* exaggerates the aura of escapism and romantic fairy-tale (typical of the operetta film), while *Der Kongreß tanzt* delivers a more plausible story.

The translations of jokes, comments, and lyrics are evidently not literal; rather, they involve adaptations, omissions, and modifications, all tailored to the respective target 'mentalities.' Therefore, when speaking of "mentality," what contemporaneous authors meant was that each version was adjusted in order to elicit a positive response from the respective audience. All these choices bet, so to speak, on the reactions and expectations of each audience, trying to empathically anticipate them.

Considered in isolation, these details might seem of little consequence, though each film produced as MVs is replete with these sorts of variations. In *The Doomed Battalion* (1932, dir. Luis Trenker), the English-language version of *Berge in Flammen/Les Monts en flammes* (1931, dir. Karl Hartl, Luis Trenker), several dialogues are extended or added in order to stress family ties, love relationships, and moments of prayer. In the German-language version, these aspects are less relevant, though not absent. I suggest that this happened because, in Germany, the film was part of a series of well-known *Bergfilme* headed by Trenker, while the anglophone and francophone universes were less familiar with this genre and needed (or, at least, were expected to need) other types of content, such as the aforementioned sentimental and religious themes. This example shows how speech was also employed on a large scale to shape the narratives transnationally, and also outside of Ufa, since this film is a production of the French Vandal & Delac.

A Mimetic Audio-Visual Translation Strategy

It is impossible to faire solche petites ici.
– Anonymous[24]

The history of MVs begins with the shooting of *Atlantic/Atlantik/Atlantis* (dir. E. A. Dupont) in 1929 – a “crucial year” for sound technologies, for avant-garde cinematographic trends, and for economic crises.[25] However, the opportunity to document MVs’ production in detail was missed, so that today we struggle to know which films were shot as MVs, which specific elements they adapted, for what reasons, and to what degree of intentionality.

Evidently, the transcultural strategy of MVs is based on the idea that each country wants to hear its own language, in line with what Benjamin calls man’s “legitimate claim to being reproduced.”[26] In other words, next to economic and political reasons, MVs were also (or especially) made for the sake of self-recognition. In relation to this, Natasa Durovicova argues that “it is this built-in duck-rabbit effect of identity politics running through the versions that makes them so worthy of study.”[27] Because of how they handled internationalization and the reproduction of films, MVs add yet another level of complexity to early sound film, complicating the relation between a film’s original and its copies.

More specifically, the conceptual problem in approaching MVs is that they essentially question the concepts of work of art, of authorship, and of stardom – in other words, they question the principle of uniqueness, instead fostering multiplicity. When each version of the ‘same’ film has its star, is there a real star at all? If only one version is successful in its national and linguistic context, can the film as *one* be considered a success?

Reflecting on intercultural and transnational issues in *Pour Vous*, film critic Wahl writes an article titled “Imitators, pay attention!”, in which he cautions the creators of “foreign versions,” or “imitators,” to be careful when imitating foreigners in productions for the international market.[28] Wahl’s choice of the term “imitators” (*imitateurs* in the original French), together

24 “Deutschland führt umbestritten in Europa.” The sentence ironically refers to the communication problems when shooting MVs with several international crews on the same set. My translation: “It is impossible to make (French ‘faire’) such (German ‘solche’) babies here (French ‘petites ici’).”

25 Bréan, Weidmann and Cornu, “Entretien avec Jean-François Cornu.”

26 Benjamin, “The Work of Art”, 232.

27 Durovicova, “Introduction,” 13.

28 Wahl, “Imitateurs, attention!”

with the "mentality transfer" described above, leads me to (re)consider MVs by focusing on the notion of *mimesis*, which has traditionally been linked to the notions of imitation and reproduction. According to this conception, there is an original reality, which is reproduced by creating a copy, which therefore always has a lower status than the original. But there is another conception, which echoes the approaches of Aristotle and Ibn Sina, and informs the work of authors such as Gilles Deleuze and Jacques Derrida, who transcend the original/copy dualism, arguing that all works have their own status of originality, while also being interdependent.[29] In this interpretation, the meaning of *mimesis* comes closer to that of "to mask, to disguise, to model, to mediate."[30] The post-structuralists have shown how the notions of mimesis and mediation are intrinsically linked, because each mimetic, imitative act is *also* one of mutual mediation between two or multiple counterparts. Derrida's works[31] on the textual domain, for instance, suggest that each text is "never the origin [...], but always doubled."[32]

Applying this to MVs, one could say that each version imitates the other versions: they are similar, but at the same time, they introduce a difference in a process of transcultural mediation and transformation, adapting to each specific cultural context. Thus, I suggest considering MVs as simultaneously *one* transnational film and *many* national(ized) versions, which address several audiences at the same time. Each version is *mimetic* in that it adapts depending on its environment – acting like a chameleon exposed to different environments, as a mediator between the work of art and the target context – through both image and sound. The reception of MVs confirms this perspective: in principle, each country privileges its own version, and the audiences look for their versions – basically, the ones they can understand (visually and linguistically). This is most evident with variations in the casts, not just in terms of their visual performances but also of their auditory ones. Like a polyhedron, then, each film shot as MVs can and should ideally be treated as a geometrical body whose unified structure is constituted by multiple interconnected faces: MVs are simultaneously one body *and* many faces.

29 Deleuze, *Différence et répétition*. Within film studies, this perspective has also been championed by Altman: in his theory of film genres, he suggests not to stop at dualisms by referring precisely to the heritage of post-structuralist criticism (Altman, "A Semantic/Syntactic Approach to Film Genre," 10).

30 Interestingly, among all Latin languages, only Italian has retained an etymological link to mimesis through the verb *mimetizzare* (meaning to camouflage).

31 Derrida, *De la grammatologie*.

32 Kelly, "Mimesis," 236.

The Emergence of a Vocal Aesthetics or "Phonogénie"

> In the opposition of the aesthetes, there is also the sting of having been bad prophets.
>
> – *La rivista cinematografica* (February 1929)[33]

Due to the communication problems on the sets, the production process of MVs was sometimes compared to the Tower of Babel. For instance, the managing director of BIP evokes this metaphor in an interview for the German press.[34] His concerns address success and audience taste: now that speech plays a role in films, will a film be as successful in the USA as in France? One possible answer comes from an article by Roger Régent and Édouard Sattler in reference to a French version of a film just shot in Hollywood: "The result is unfortunate because, despite their good will, these actors reveal an accent in the talkie, often deplorable, that threatens to compromise the success of the film."[35]

Issues related to voice, pronunciation, and accents indeed become serious risks for the careers of film stars, such as for John Gilbert, who failed the "microphone test."[36] With the talkies, voices become important acoustic components of films, to the point that a new concept emerges, especially in the French context: in addition to being photogenic, film stars are now judged for being *phono*genic (or not): for instance, Marlene Dietrich is praised for her rough, entrancing voice, Lilian Harvey for her joyful, playful vocal touch, Albert Préjean for his honest voice, Henry Garat for his tender tonalities.[37] Also, next to the already common surveys for choosing the most beautiful actor or actress, surveys about the best screen voice appear, such as the prize for "the most phonogenic star" awarded in *Pour Vous* in 1931.[38]

The specific qualities of voices had consequences for the aesthetics of films in general, with MVs adding another level of complexity, operating transculturally and often on several versions at once. This is evident in the Ufa film *F.P.1 antwortet nicht/I.F.1 ne répond pas/F.P.1* (1932, dir. Karl Hartl)

33 R., "L'arte muta è minacciata dai progressi del fonocinematografo?"

34 Aros, "Der Turmbau zu Babel."

35 Régent and Sattler, "Vernon via Séville, Fédor par le Mexique."

36 The fact that his voice was not phonogenic was widely discussed in the press. In *Pour Vous*, for instance, readers and writers comment on how his voice was too soft and high-pitched to fit his virile, powerful physical presence (Genova, "La carrière de John Gilbert est-elle en danger?").

37 Doré, "Devant le micro." All stars mentioned here were involved in the production of MVs.

38 Anonymous, "Voici les lauréats du jeu de la vedette la plus photogénique ..."

Figure 12.2: Collector card from *F.P.1 antwortet nicht/I.F.1 ne répond pas/F.P.1* (*Floating Platform 1*) (1932, dir. Karl Hartl) showing Hans Albers, Charles Boyer, and Conrad Veidt in the role of Ellissen.

starring Hans Albers, Charles Boyer, and Conrad Veidt, respectively, in the 'same' role of pilot Ellissen (Fig. 12.2). This film is a hymn to technology and technical progress, similar to *Die Nacht gehört uns/La nuit est à nous* (*The Night belongs to Us*) (1929, dir. Carl Froelich, Henry Roussell), which also stars Albers in the German version. Most communications happen through the machines placed on the F.P.1, a "floating platform" built in the middle of the Atlantic Ocean,[39] so that the film is a continual audio-visual display of telephones, radios, microphones, etc.

The composition of the casts, their performances, and national reception(s) are worthy of attention here: Albers in the German version plays a globetrotter, a rebellious, sensual Don Giovanni; he has a strong presence on the screen, not least due to his voice, which even in contemporary sources is still praised as beautifully intense and named as a sure reason for his success.[40] Boyer's French Ellissen sounds less self-assured, though his vocal ability was generally praised: "His famous voice, which exerts such a powerful charm on all audiences, is like a precision instrument of extreme sensitivity, which

39 The platform was actually built near the German island of Rügen, where the shooting took place.

40 As in the commentary in the documentary "Die Liebe des Hans Albers" (NDR, January 2021). On the reception of Albers's voice in this film, see also Wiegand, "'Eine Entdeckungsfahrt in die Welt der Geräusche.'"

he uses with perfectly natural art and delicacy."[41] In the English version, the Ellissen played by German Conrad Veidt is more contained, praised for his charm of a gentleman; also, due to his previous successes in the UK, Veidt was deliberately chosen by Ufa to perform in English, since the British audience was already accustomed to him. Furthermore, the costumes and the physical features of the three Ellissens look different – in contrast to the three similar-looking actors playing Pepi/Pépi (as seen in Fig. 12.1), as well as to Fritsch and Garat (as the Tzar) in *Der Kongreß tanzt/Le congrès s'amuse/The Congress Dances*. This shows that approaches to MVs sometimes differed even within the same studio (in this case, Ufa).

As for the actresses playing the character of Claire, Sybille Schmitz in *F.P.1 antwortet nicht* and Jill Esmond in *F.P.1*, they display differences in their ways of speaking, acting, and dressing, too: Esmond has the charm of a serious, sober woman, while Schmitz is much more provocative and sensual (after this film, she would become a model of the confident and independent New Woman in Germany). Finally, the screenplay of *F.P.1 antwortet nicht* contains several notes saying "only for the German version," especially in reference to songs and lyrics, suggesting that some variations were deliberately planned for certain versions *exclusively*.[42]

Given all these details, it appears that the three versions were planned to be different on purpose, which shaped the actors' performances – from gestures to voices and emotional reactions – in line with the idea of mirroring cultural nuances and expectations to make each version more appealing in its context.[43] In addition, the choice of the actors appears to have been driven by differing, nationally specific ideas of what a phonogenic voice is, as also mirrored in the films' reception.

Conclusion

MVs were conceived and constructed as a flexible adaptation process of nationally and culturally specific elements, which makes them a privileged field for the analysis of cultural transfers – regardless of the actual reasons behind the variations and of the films' success.

41 Thomasson, "Après trois ans d'absence, Charles Boyer revient parmi nous ses projets."

42 Asper, "'The Top Picture to Top All Pictures,'" 127 and 133.

43 Similar dynamics in relation to different clothing styles, attitudes, and expressions can be found e.g., in *Der Sohn der weißen Berge/Les Chevaliers de la Montagne* (*The Son of the White Mountain*) (1930, dir. Luis Trenker, Mario Bonnard).

In this article, I have investigated why MVs were promoted as a solution to the challenges of internationalism and linguistic diversity in Europe, showing the implications of their multiplicity and their influence on the evolution of film aesthetics. In light of language barriers, MVs offered an opportunity to experiment with the potentials of sound films, to transfer them mimetically to other countries, and to adapt them on the national and cultural scale. To approach the phenomenon *tout court*, I have suggested to look at MVs as a mimetic technique, insofar as each version has its own status of originality and its own impact in its respective target context. One could argue that the versions shot in the language of the production country should be considered the 'originals,' since they were often privileged by the producers and more successful than the others – an aspect that requires further study. However, a look at the films' reception reveals that national audiences considered *their* respective version the main – or even the only – film, above all because they were exposed to that version exclusively: what was seen in the UK and in the USA was not the emotional, rebellious Ellissen of *F.P.1 antwortet nicht* but the contained, gentlemanly Ellissen of *F.P.1*; the French audience was exposed to the jokes in *Le congrès s'amuse* but not to those of *Der Kongreß tanzt*, and so on.[44]

What I especially want to stress here is the major contribution of MVs to (sound) film history: that is, the integration of culturally and nationally oriented elements primarily through actors and actresses as vehicles of language, voice, and cultural signification. This had a major impact on the global cinematic experience, because audiences of different countries were exposed to the 'same' film, yet through transculturally mediated versions of that film. MVs are indeed a unique audio-visual translation technique that operates by transcultural mediation.

Finally, MVs remained a short-lived 'attempt' because they quickly faded with the consolidation of dubbing – a faster and more affordable strategy – by late 1932. As mentioned earlier, though, MVs never totally disappeared from the universe of cinema. In recent years, the practice even crossed continental borders.[45] This makes it even more important to fully map the networks of MVs at the beginning of (sound) film history and to examine the role of this mimetic technique with its multiple levels of influence.

44 Only in the capitals and in cities along the Franco-German border (e.g., Strasbourg) was it possible to see both versions, usually only for a very short time. Thus, for most audiences, it was almost impossible to see several versions of the same film.

45 See footnote 14.

Bibliography

Altman, Rick. "Penser l'histoire du cinéma autrement: un modèle de crise." *Vingtième Siècle, revue d'histoire* 46 (1995): 65–74.

Altman, Rick. "A Semantic/Syntactic Approach to Film Genre." *Cinema Journal* 23, no. 3 (1984): 6–18.

Arnoux, Alexandre. "Il peut se créer un art neuf." *Pour Vous*, November 29, 1928.

Arnoux, Alexandre. "La bataille des langages et le film parlant." *Pour Vous*, September 5, 1929.

Aros. "Der Turmbau zu Babel: Ein Interview mit Generaldirektor Dent aus London." *Der Kinematograph*, May 21, 1931.

Asper, Helmut G. "'The Top Picture to Top All Pictures': Walter Reischs Drehbuch für *F.P.1 antwortet nicht*." In *Walter Reisch. Film schreiben*, edited by Günter Kreen, 120–37. Wien: Filmarchiv Austria, 2004.

Barnier, Martin. "Versions multiples et langues en Europe." *Mise au Point*, no. 5 (2013). Accessed March 5, 2023, https://journals.openedition.org/map/1490.

Benjamin, Walter. "The Work of Art in the Age of Mechanical Reproduction" [1936]. In *Illuminations*. Translated by Harry Zohn, 217–51. New York: Schocken Books, 1969.

Bréan, Samuel, Anne-Lise Weidmann and Jean-François Cornu. "Entretien avec Jean-François Cornu." *Traduire* 230 (2014): 10–21.

Chion, Michel. *The Voice in Cinema*. Translated by Claudia Gorbman. New York: Columbia University Press, 1999.

Claus, Horst and Anne Jäckel. "*Der Kongreß tanzt*: Revisited." *CINEMA & Cie*, no. 6 (2005): 76–95.

Deleuze, Gilles. *Différence et répétition*. Paris: Presses Universitaires de France, 1968.

Derrida, Jacques. *De la grammatologie*. Paris: Les Éditions de Minuit, 1967.

"Deutschland führt unbestritten in Europa." *Der Kinematograph*, October 6, 1931.

Doré, Claude. "Devant le micro." *Ciné-Miroir*, February 26, 1932.

Durovicova, Natasa. "Introduction." *CINEMA & Cie*, no. 4 (2004): 7–16.

Elsaesser, Thomas. "ImpersoNations: National Cinema, Historical Imaginaries and New Cinema Europe." *Mise au Point*, no. 5 (2013), https://journals.openedition.org/map/1480.

Garncarz, Joseph. "Making Films Comprehensible and Popular Abroad: The Innovative Strategy of Multiple-language Versions." *CINEMA & Cie*, no. 4 (2004): 72–79.

Garncarz, Joseph. *Wechselnde Vorlieben. Über die Filmpräferenzen der Europäer 1896–1939*. Frankfurt am Main: Stroemfeld, 2015.

Genova. "La carrière de John Gilbert est-elle en danger?" *Ciné-Miroir*, March 7, 1930.

Hardt, Ursula. *From Caligari to California: Erich Pommer's Life in the International Film Wars*. New York: Berghahn, 1996.

"Il film internazionale finirà?" *La rivista cinematografica*, January 1, 1930.
Kelly, Michael. "Mimesis." *The Encyclopedia of Aesthetics*. Vol. 3. New York: Oxford University Press, 1998.
Klimes, Ivan and Pavel Zeman. "Aktualita 1937–1938." *CINEMA & Cie*, no. 6 (2005): 62–70.
Klitzsch, Ludwig. "Der Film als nationaler Kulturfaktor." *Der Kinematograph*, July 20, 1932.
Natanson. "Les joies et les ennuis du dialogueur." *Pour Vous*, October 13, 1932.
Pommer, Erich. "Pommer über den mehrsprachigen Tonfilm." *Der Kinematograph*, September 17, 1930.
R., O. "L'arte muta è minacciata dai progressi del fonocinematografo?" *La rivista cinematografica*, February 28, 1929.
Régent, Roger and Édouard Sattler. "France, Allemagne, Italie …" *Pour Vous*, September 25, 1930, 6.
Régent, Roger and Édouard Sattler. "Vernon via Séville, Fédor par le Mexique." *Pour Vous*, February 19, 1931.
Szczepanik, Petr. "The 'Birth' of Cinema History: The Coming of Sound Films and Media's Invention of Their Past." In: *MLVs, Cinema and Other Media*, edited by Veronica Innocenti, 29–43. Udine: Campanotto Editore, 2006.
Thomasson, R. "Après trois ans d'absence, Charles Boyer revient parmi nous. Ses projets." *Pour Vous*, March 2, 1933.
"Voici les lauréats du jeu de la vedette la plus photogénique …" *Pour Vous*, December 17, 1931.
Wahl, Lucien. "Imitateurs, attention!" *Pour Vous*, July 3, 1930.
Wiegand, Daniel. "'Eine Entdeckungsfahrt in die Welt der Geräusche': Zur historischen Rezeption des frühen Tonfilms *Die Nacht gehört uns* (D 1929)." *Montage AV* 2 (2021): 59–75.

About the Author

Maria Adorno is a PhD candidate in film history at the University of Cologne in cooperation with the Universities of Udine and Lyon 2. Her research focuses on the transcultural dimension of multiple versions in Europe during the early sound film period. Recent articles include "Cinéma national ou international? Le débat franco-allemand sur le film parlant européen à travers la presse et les versions multiples (1928–1932)," in *Revue d'Allemagne et des pays de langue allemande* (2022). She is co-director of the *Karlsruhe Silent Film Festival* and contributor for the website *Sinn und Cinema*.

V

Music and Noise

13. "How Did the Music Get to the Fish Market?"

On the Use of Nondiegetic Music in Early German Sound Films

Daniel Wiegand

Abstract: This chapter traces patterns in the use of nondiegetic music between late 1929 and mid-1931 based on the analysis of a significant corpus of early German sound films. While nondiegetic music did occur during this period, it was mostly limited to certain film genres and standard moments. More exceptional uses emerged in individual films before becoming more frequent from mid-1931 on. In the second part, the chapter shows that the initial reluctance to use nondiegetic music was partly supported by authors in the trade press since it was often considered to be 'illogical,' 'arbitrary,' or incompatible with sound film's realist (or 'objective') aesthetics. However, this disapproval gradually gave way to habituation and fascination when new ways of connecting images with music emerged.

Keywords: film music, musical score, historical reception, genre, realism

Histories of film music often tell us that scoring practices[1] in American sound film only fully emerged in the so-called 'classical period,' initiated by pioneering films like *King Kong* (1933, dir. Merian C. Cooper/Ernest B. Schoedsack). Nondiegetic music, it seems, was rarely used in the first years of the sound era.[2] In his illuminating study of the Hollywood film score in the transitional era, Michael Slowik partly refutes these claims. He takes

1 In this chapter, I use "nondiegetic music," "(musical) score," and "scoring" mostly as synonyms, except in a few cases where a differentiation seems appropriate.
2 See e.g., Cooke, *A History of Film Music*, 85.

Wiegand, D. (ed.), *Aesthetics of Early Sound Film: Media Change around 1930*. Taylor & Francis Group, 2023
DOI 10.5117/9789463727372_CH13

a closer look at a large corpus of films from the period and finds that "the early sound era [...] featured a wide *array* of musical approaches rather than single-minded avoidance of nondiegetic music."[3] Slowik does confirm, however, that a tendency toward a "sparse musical style" existed in the years between 1928 and 1931, which he finds notable, given the general practice of *continuous scoring* during the silent era as well as in synchronized films and part-talkies at the end of the period.[4]

Research about the specific uses of nondiegetic music in early German sound films is scarce. Often, films like *M* (1931, dir. Fritz Lang) or *Der blaue Engel* (*The Blue Angel*) (1930, dir. Josef von Sternberg) are mentioned as examples of deliberate attempts to do without much scoring and to even leave several scenes and moments completely silent. But how exceptional were these films? What were the standard practices? When, how, and why did they change? It seems crucial to tackle these questions in greater detail to develop a more comprehensive understanding of how audio-visual film style in Germany was transformed during the transition to sound.

In this chapter, I argue that scoring practices for sound films existed in Germany from the beginning, but they were quickly confined to a fixed and rather limited set of standard uses and specific film genres. Outside of these confines, which persisted from late 1929 to mid-1931, nondiegetic music was used only in a few isolated instances and in a small number of daring films that were hotly debated at the time. However, the scoring of individual scenes, including dialogue underscoring, became more common across genres in the second half of 1931. My argument is based on a sample of more than sixty feature-length fiction films with sound, all of them full talkies, produced and released in Germany between late 1929 and mid-1931.[5] Since many of these films have never been released on DVD, I mostly viewed them as digital copies obtained from collectors' circles. In

3 Slowik, *After the Silents*, 2.

4 Ibid., 87–135. Likewise, Charles O'Brien argues that "the reduction [...] of the amount of musical accompaniment" is a key difference between late silent film and early sound film (referring to both diegetic and nondiegetic music). O'Brien's analysis of a larger corpus of American films shows that in 1929–1930, "more than 50 percent of the average feature film's running time was music-free," with the figure climbing to over 70 per cent by 1931 (O'Brien, *Movies, Songs, and Electric Sounds*, 53–54).

5 This equals roughly 25 per cent of Germany's sound feature film output during that time and almost 30 per cent for 1930 alone. The exact number is difficult to ascertain and depends on which films one counts. The percentages given here are based on the lists in Bauer 1950 and Klaus 1989, which also include hybrid films such as part-talkies, foreign productions of German-language films, Austrian productions screened in Germany, and films never released, as well as some non-fiction features and short films, all of which are excluded from the present corpus. Thus, based on the

order to trace the stylistic changes during 1931, I analysed a smaller sample of eighteen films from the second half of that year, which produces results significant enough to buttress the argument.

Part-Talkies and Synchronized Films – A Blind Spot in German Film History

Like in Hollywood, the first feature-length sound films produced in Germany were synchronized films and part-talkies. Synchronized films, such as Walter Ruttmann's *Melodie der Welt* (*Melody of the World*) (March 1929), mostly featured a continuous score with occasional sound effects; part-talkies integrated songs and audible dialogues only in selected scenes.[6] We know that Hollywood's part-talkies typically continued the score from the 'silent' parts into the talking sequences.[7] Even though it is difficult to make a comparable statement about German part-talkies (most of them have to be considered lost), contemporary reviews indicate that at least some of them underscored dialogue parts as well. This also seems likely because Hollywood's part-talkie *The Singing Fool* (1928, dir. Alan Crosland) was one of the first sound films successfully screened in Germany, and its continuous dialogue underscoring may have served as a model.[8] A frequent use of the score in the first German part-talkies would also explain why Germany's first *full talkie*, the musical melodrama *Melodie des Herzens* (*Melody of the Heart*) (December 1929, dir. Hanns Schwarz), still contains a lot of scoring, even for some of the dialogue sequences. However, as the following section will show, the use of a musical score became much less common in subsequent all-talking films.

'Heavy Scoring'

None of the films in the sample continued the practice of full scoring known from silent films, and only 15 made use of what I call *heavy scoring*, that

corpus used here, the percentage is probably even higher. The release dates in the corpus range from December 1929 (when the first German full talkies were released) to the end of June 1931.

6 Wiegand, "'Islands of Sound.'"

7 Slowik, *After the Silents*, 40–86.

8 It seems that later part-talkies were more reluctant. The surviving *Cyankali* (*Cyanide*) (1930, dir. Hans Tintner) features only one dialogue scene at the end, which is unscored (for a close reading of this film, see Wiegand, "'Islands of Sound.'")

is, the use of nondiegetic music for more than just a few brief sequences. The majority of these films were musical films (10), most notably operetta films by Ufa (4). This is not surprising given that for Wilhelm Thiele, the 'inventor' of the subgenre, the free use of nondiegetic music was a defining quality of these films. As he writes in a short 'manifesto' shortly before the release of his *Liebeswalzer* (*The Love Waltz*) (1930), which initiated the series: "The music has to drive the visuals forward […] without optical or musical breaks." Hence, while "in films with a serious subject, the story and the sound have to come from the same source, this does not have to be the case in the operetta film."[9] Correspondingly, nondiegetic music in operetta films by Thiele and others would often be used to accompany songs, dances, dialogues, or characters' rhythmic movements, deploying devices such as mickey-mousing and abrupt shifts from speaking to singing and vice versa.[10]

Considering the popularity of musicals in early German sound film, the comparatively low number of 15 films here indicates that most musicals (including other subgenres such as the backstage musical and the 'singer film') only deployed nondiegetic music very reluctantly and mostly relied on *diegetic* music to accompany songs.

The second distinct genre using heavy scoring was the mountain film. Apparently, the mountain scenery, which served as the setting for these sensational melodramas, justified the use of nondiegetic music, including dialogue underscoring. It seems that only two mountain films were produced until mid-1931 – *Der Sohn der weißen Berge* (*The Son of the White Mountain*) (1930, dir. Mario Bonnard) and *Stürme über dem Montblanc* (*Storm over Mont Blanc*) (1930, dir. Arnold Fanck) – but many later examples used the same kind of heavy scoring.

Apart from musicals and mountain films, only a handful of exceptional films in the sample made use of heavy scoring. The adventure film *Hans in allen Gassen* (*Hans in Every Street*) (1930, dir. Carl Froelich) features several scored montage and transitional sequences but interrupts the score as

9 Thiele, "Operette im Tonfilm."

10 In the narrow definition proposed by Michael Wedel, only Thiele's *Der Liebeswalzer* and *Die Drei von der Tankstelle* (*The Three from the Filling Station*) (both 1930) as well as Hanns Schwarz's *Einbrecher* (*Burglars*) (1930) and *Ihre Hoheit befiehlt* (*Her Grace Commands*) fall into the category of the operetta film until mid-1931 (Wedel, *Der Deutsche Musikfilm*, 271). The other musical films in the sample that use heavy scoring – more or less borrowing from the style of Ufa's operetta films – are *Die vom Rummelplatz* (*Fair People*) (1930, dir. Karl Lamac), *Die zärtlichen Verwandten* (*Affectionate Relatives*) (1930, dir. Richard Oswald), *Eine Stunde Glück* (*One Hour of Happiness*) (1931, dir. Wilhelm Dieterle), *Die 3 Groschen-Oper* (*The 3 Penny Opera*) (1931, dir. G. W. Pabst), and Ufa's early effort *Melodie des Herzens*, discussed above.

soon as dialogue sets in. More daringly, Ufa's crime comedy *Der Mann, der seinen Mörder sucht* (*Looking for His Murderer*) (1931, dir. Robert Siodmak) uses nondiegetic music to accompany character movements in the manner developed by the company's operetta films. Next, the historical melodrama *Der Mörder Dimitri Karamasoff* (*The Murderer Dimitri Karamazov*) (1931, dir. Fedor Ozep) uses a symphonic score for several romantic scenes as well as for underscoring dialogues. The film was clearly devised and perceived as an experiment, and its exceptional degree of scoring was widely discussed in the trade press.[11] Finally, *Das Kabinett des Dr. Larifari* (*The Cabinet of Dr. Larifari*) (1930, dir. Robert Wohlmuth) is already a sound film parody, which includes scored operetta and mountain scenes.

This brief overview allows us to draw initial conclusions. Even though broadening the corpus might uncover some more unusual cases, the sample indicates that heavy scoring was not widely used at all outside the genres of the musical and the mountain film. Even *within* the musical genre, most films that did not deploy the operetta formula refrained from it. Thus, while technological limitations – for instance concerning the possibility for rerecording – certainly played a role in the reluctant use of the score as well,[12] the operettas and mountain films show that heavy scoring, including dialogue underscoring, was perfectly possible if desired and that reasons for *not* using it must have included other factors as well. It seems that aesthetic considerations, especially when they pertained to questions of genre, seemed to be a decisive factor here. At the same time, the few examples of heavy scoring in other genres, most notably in *Der Mörder Dimitri Karamasoff*, reveal that even deviations from genre conventions were possible and that some filmmakers (like the Soviet immigrant Fedor Ozep) were willing to stretch the boundaries of convention early on.

Sparing uses

This section discusses how nondiegetic music occurs in the remaining 45 films in the sample, which are not 'heavily scored.' Here, I differentiate between *standard uses of scoring* and more *exceptional uses*. By far the most common practice was the *scoring of the opening credits* (or, in rare cases, of

11 Evaluations were mostly positive. See e.g., Jäger, "Filmkritik: Der Mörder Dimitri Karamasoff."

12 On rerecording and other ways of sound mixing in early German sound film, see Müller, *Vom Stummfilm zum Tonfilm*, 214–21. On rerecording in early Hollywood sound film, see Jacobs, "The Innovation of Re-Recording."

blank film if there were no credits). From a later historical perspective, we might expect this to occur in virtually every film; however, this is by no means the case. Unlike Hollywood films of the period, several early German sound films (9 in the sample)[13] left the opening credits either completely silent (except for an occasional announcing bell like in *M*) or had the credits spoken by one or several announcers rather than using music (which became something like a signature style in the films produced and directed by Richard Oswald). Silent opening credits were typical of the films from the production company Nero but can also be found in the crime movies *Täter gesucht* (*Perpetrator Wanted*) (1931, dir. Carl Heinz Wolff) and *D-Zug 13 hat Verspätung* (*Express 13*) (1931, dir. Alfred Zeisler).[14]

The scoring of the opening credits would sometimes continue into the first scene, which is what I call the *opening credits overlap* (8). In some cases, the same tune from the opening credits would simply continue for a few seconds into the first shot, in others, a new theme would set in to emphasize a striking shot or a brief sequence of shots. In *Der Tiger* (*The Tiger Murder Case*) (1930, dir. Johannes Meyer), the opening score turns dramatic for the first shot, in which we see a murder victim on the nightly streets; in *Die Nacht gehört uns* (*The Night Belongs to Us*) (1929, dir. Carl Froelich), the final chord of the opening theme grows into a dramatic fanfare when the headlights of a car approach the camera, and the music continues for a few sombre bars while we read a newspaper note on an upcoming car race. Occasionally, the newly starting theme was even extended over a longer sequence, such as a montage of shots of the port of Genova in *Er oder Ich* (*He or I*) (1930, dir. Harry Piel). Sometimes, the opening credits overlap was used for little gags that momentarily cast doubt on the music's status. This was mostly done in musical films. In *Das Lied ist aus* (*The Song is Ended*) (1930, dir. Géza von Bolváry), the score from the credits continues over a close-up of a hand, which belongs to someone at a party knocking against a glass of champagne in order to deliver a speech; as an allusion to the film's title, the music stops. Has it been diegetic music all along? In *Ein Tango für dich* (*A Tango for You*), released a few months earlier in 1930, director von Bolváry takes these ambiguities even further. The score leads into the first shot of a corridor in a girls' boarding school, where some of the girls are sneaking towards a friend's room. Once they enter the room, the music stops and

13 With two films, the case is unclear, because the opening credits were missing from the copies used.

14 Some critics even complained when otherwise 'objective' crime films *did* have a score during the opening credits (e.g., "Kritische Filmschau: Der Schuss im Tonfilmatelier," 6).

one of the girls says: "We heard music." As it turns out, the girl, Mady, had a phonograph hidden in her bed; however, the other girls keep referring to a "saxophone" in the music, which wasn't part of the opening score, and when Mady plays the record again, the song is clearly a different one.

As these examples show, the use of the opening credits overlap was by no means conventionalized in this early period. It is also notable that the vast majority of films in the corpus do *not* use an overlap and instead stop the music at the end of the opening credits.

Another standard practice was the scoring of the film ending, which I call the *final score*. This can encompass the entire last scene of the film, just the last seconds of it, or nothing but the end title card (German films from the early sound period generally do not have full credits at the end). Notable examples with extended scoring during the final scene/shot include *Der blaue Engel* and *Schuberts Frühlingstraum* (*Schubert's Dream of Spring*) (1931, dir. Richard Oswald). However, just as with beginnings, the final score occurs less often than might be expected. From the available film prints it seems that several films (13) even end without any kind of music, either in complete silence or with a simple sound effect (like the stuttering car engines in *Die Nacht gehört uns*). These films mostly belong to the 'serious' drama genre, but there are also examples in lighter films and comedies (e.g., *Ihre Majestät die Liebe*, *Her Majesty Love*, 1931, dir. Joe May), and even in heavily scored operetta films (*Eine Stunde Glück*).[15] If there *was* music at the end, it was sometimes diegetic, especially when it had been present in the final scene anyway and could simply be extended over the end title card (sometimes with a slight change in quality to signify a 'shift' towards the nondiegetic). Whether one includes these cases or not, the overlap occurs much more often with the final score than at the films' beginning.[16]

The last standard use was the *scoring of landscape shots*, typically used as a transition between scenes or as an illustration of a journey. Often, these shots showed well-known holiday destinations in Italy or Austria, sometimes particular cities such as Venice or Istanbul. This kind of soring occurs in eight of the films in the corpus, and only once or twice in each film (in contrast to the heavy scoring in the mountain films discussed above). A special case is the rather late *Salto mortale* (*Trapeze*) (May 1931, dir. E. A.

15 It should be noted, however, that it is not always clear whether the surviving print is complete. Silent endings and missing title cards in early German sound film would thus merit in-depth research in their own right.

16 One reason might be that end titles were generally much shorter than opening credits, and an overlap would thus allow the music to play for more than just a few seconds.

Dupont),[17] in which landscape scoring sets in during the previous scene, where it melodramatically underlines the female protagonist's sadness before it continues over shots of dull winter landscapes. In contrast to earlier films, the landscapes here are not recognizable tourist destinations but evoke an atmosphere of desolation. This decidedly melodramatic use of the score anticipates developments that would emerge more forcefully during the next months.

Apart from these standard uses, there were more *exceptional uses*. Only three out of the 45 films use a score for *montage sequences* that went beyond the landscape formula outlined above (with *Salto mortale* being a borderline case). Towards the end of *Boykott* (*Boycott*) (1930, dir. Robert Land), shots of the young protagonist walking through the streets of Berlin are accompanied by a dramatic score; a rapid montage of rushing trains and railroad lines superimposed over his face illustrate his perturbed state of mind, suggesting that he is contemplating suicide (Fig. 13.1). The sequence stands out from what is otherwise a very quiet and sombre film and recalls the famous scene when Alice has killed the rapist in *Blackmail* (1929), Alfred Hitchcock's first sound film, which had been screened in Germany as well (and thus may have served as a model for *Boykott*). In *Brand in der Oper* (*Fire in the Opera House*) (1930, dir. Carl Froelich), an even more sophisticated and dramatically scored montage sequence illustrates the female protagonist's feverish dreams as she is recuperating from the eponymous fire. The sequence employs blurred images, low-key lighting, negative printing, multiple exposures, and other means to distort the images. Finally, in *Abschied* (*Farewell*) (1930, dir. Robert Siodmak), a slower and more lyrical montage of deserted objects and corners in a tenement illustrates the passage of time and is scored by the film's main theme (a love song), familiar from the opening credits. While *Abschied* makes use of a *diegetic* piano throughout, played by one of the lodgers and heard everywhere in the building, this is the only moment when it is complemented by an orchestra, which obviously cannot be present in the lower-class apartment building. Such discreet shifts from diegetic to nondiegetic – Michael Slowik terms them "diegetic withdrawals" – were rare in German films at the time.[18]

Even less frequent than the scoring of montage sequences was the scoring of *showdowns and action sequences*. In the films without heavy scoring, such scenes were almost always left unscored, in contrast to later standard

17 The film had its German premiere only in the second half of 1931, but the world premiere took place in Vienna in May (Klaus, *Deutsche Tonfilme, vol.* 2, 243–44).

18 Slowik, *After the Silents*, 112–16.

Figure 13.1: A rare moment of nondiegetic music in an early German sound film, occurring during a montage sequence in *Boykott* (*Boycott*, 1930, dir. Robert Land) (courtesy of Deutsche Kinemathek).

Figure 13.2: A chase sequence in *Das Flötenkonzert von Sanssouci (The Flute Concert of Sans-Souci)* (1930, dir. Gustav Ucicky), left completey unscored (courtesy of Friedrich-Wilhelm-Murnau-Stiftung).

practice. For instance, the historical drama *Das Flötenkonzert von Sanssouci* (*The Flute Concert of Sans-Souci*) (1930, dir. Gustav Ucicky) contains a chase on horseback that lasts for several minutes with nothing but the sound of the hooves on the soundtrack (Fig. 13.2). Similarly, *Er oder Ich*, the first sound film by action pioneer Harry Piel, and *Brand in der Oper* contain dramatic action sequences left completely unscored.[19] The only exception in the sample is, again, *Salto mortale*, which offers another instance of "diegetic withdrawal." Here, the music heard during the circus shows sometimes seems slightly inappropriate for a diegetic circus orchestra (especially during the trapeze sequences), and it clearly turns nondiegetic when one of the acrobats is in serious danger of falling to the ground – a moment when a diegetic orchestra would certainly have stopped playing. Once again, *Salto mortale*, released at the very end of the period discussed here, proves to be a film that explores new ground.

Dialogue underscoring does appear in several films but is usually marked as diegetic (again, this is different in the films using 'heavy scoring'). Only rarely does the music's status remain somewhat ambiguous. In *Brand in der Oper*, an underscored dialogue takes place on the terrace of a café so that the music could potentially come from inside, even though it seems a little too loud for this. Similarly, the Ufa production *Das gestohlene Gesicht* (*The Stolen Face*) (1930, dir. Philipp Lothar Mayring/Erich Schmidt) features two scenes in a hotel lobby, where the music might originate from a band playing in some other room, even though we don't see any band. While these scenes leave the 'diegetic option' open, there are no examples in this corpus of 45 films of dialogue underscoring that is unambiguously nondiegetic.

Finally, there is the isolated example of what one could call a 'commentary function' of music in the historical drama *Danton* (1931, dir. Hans Behrendt), where nondiegetic military drums and fanfares (rather than a true 'score') accompany key historical moments of the French Revolution.

As these examples show, there was indeed a wide range of different uses of nondiegetic music in the period up until mid-1931. Directors and producers were certainly testing out several options. However, as the sample also indicates, nondiegetic music was mostly limited to specific genres and a few clearly demarcated standard uses, with relatively few exceptional cases.

19 Tellingly, these films use a score for a landscape montage and a dream sequence, respectively, indicating that these types of sequences were more likely to be scored.

Toward a More Saturated Musical Soundtrack

The situation begins to change in mid-1931. Exceptional uses of scoring, such as scoring of montage sequences, become more frequent, and functions of the score that were rarely explored before, such as the scoring of dramatic high points or romantic scenes, become common across all genres. One example is the highly emotional scene of the female teacher kissing Manuela good-bye in *Mädchen in Uniform* (*Girls in Uniform*) (November 1931, dir. Leontine Sagan), a film that also features several scored scenes depicting the activities of girls at a boarding school. At the same time, other key scenes in this film remain without a score, even when there are no dialogues or sound effects, as during the suicide attempt and the silent ending. Another example of this *selective approach to scoring* is *24 Stunden aus dem Leben einer Frau* (*24 Hours in the Life of a Woman*) (1931, dir. Robert Land), which combines standard scoring for the opening credits, end titles, and transitional landscape/travelling sequences with more daring uses, such as scoring for the introduction of Henny Porten's main character (though still in combination with landscape shots) and for the film's dramatic climax (the casino scene towards the end). Nevertheless, other scenes are mostly silent, even some of the landscape and travelling shots, which still seems unusual from the perspective of later 'classical' scoring practices.

The use of nondiegetic music for montage sequences and for depicting subjective states of mind became more common as well, as in *Der Raub der Mona Lisa* (*The Theft of the Mona Lisa*) (August 1931, dir. Géza von Bolváry), *Berlin – Alexanderplatz* (October 1931, dir. Piel Jutzi), or *Der brave Sünder* (*The Upright Sinner*) (October 1931, dir. Fritz Kortner), which also feature repeated cases of "diegetic withdrawal."[20] Mountain films such as *Der weisse Rausch* (*White Frenzy*) (Dec. 1931, dir. Arnold Fanck) continued to rely on heavy scoring, especially in outdoor scenes. Simultaneously, films with hardly any nondiegetic music were still made in late 1931, among them the literary adaption *Der Hauptmann von Köpenick* (*The Captain from Köpenick*) (dir. Richard Oswald) and crime movies such as *Der Draufgänger* (*The Go-getter*) (dir. Richard Eichberg) or *Der Zinker* (*The Informer*) (dir. Karl Lamac).[21] Hence, rather than a clear shift towards heavy scoring, the second half of

20 For a discussion of the dream sequence in *Der brave Sünder*, see Katharina Loew's chapter in this volume.

21 That the lack of a score in *Der Zinker* is conspicuous and highly unusual from a later perspective can be seen from the fact that a new continuous score was composed by Florian C. Reithner for the German DVD edition in 2009 (Koch Media).

1931 saw a slight but notable *increase* as well as a greater *diversification* of scoring practices compared to the earlier period.

Reservations against Nondiegetic Music in the Trade Press

During the early sound film era in Germany, spectators – and especially professional writers – were highly sensitive to the way that speech, music, and noises were utilized in specific films, and they often put forward normative rules for their uses. This was true from the very beginning of the conversion, when only a few short films with sound were produced, and knowledge about longer films often came exclusively from reports from the US or the UK. At this time, writers often assigned the three registers of sound (speech, music, noises) to what they considered distinct *types* of sound film, which should ideally be kept separate from each other: the talkie, the musical film, and the "noise film." While some argued for a strict separation of these three forms, others reasoned about possible mixtures.[22] Hence, when the first feature-length sound films were released in Germany, the use of music, especially nondiegetic music, was widely debated in reviews and in articles by film critics, filmmakers, and composers. As I will argue in the remainder of this chapter, the initial reluctance to use nondiegetic music that we find in the films was partly reflected in these writings. I will discuss four different but intertwining arguments that were prevalent at the time.

(1) The 'intelligibility argument': *Dialogue underscoring renders dialogues unintelligible.* This argument has sometimes been assumed to play a major role in the reluctant use of nondiegetic music in early sound films. Indeed, one can find several German reviews that complain about underscoring on these grounds. Since underscoring rarely occurred in full talkies made in Germany, most of these statements relate to part-talkies or foreign productions. For instance, an anonymous reviewer of MGM's German-language production *Die Königsloge* (*The Royal Box*) (USA 1929, dir. Bryan Foy) referred to the film's score as a "disturbing noise that rendered the words, which were barely intelligible anyway, even less intelligible."[23] However, the argument seems to pop up only occasionally, probably also due to shifting qualities in sound reproduction in the respective cinemas. Moreover, while poor intelligibility would only speak against the specific practice of dialogue

22 See e.g., Chaparral, "Dernier cri in Hollywood" and Schacht, "Tonfilmdramaturgie."

23 "Kritische Filmschau: Die Königsloge," 8.

underscoring, most of the criticism in the trade press was aimed at scoring and nondiegetic music in general.

(2) The 'logic argument': *Nondiegetic music is illogical and confusing when it is mixed with diegetic sound.* There is no shortage of historical accounts featuring anecdotes about audience members who were wondering where the music was coming from when it did not originate in the film's story world. Looking back, American film composer Max Steiner stated in 1937: "You had to have an orchestra on view, or a phonograph or performance, so that people would not wonder where the music was coming from."[24] Whether or not individual spectators were actually confused is difficult to ascertain, but there are indeed many comments in the German trade press stating that nondiegetic music contradicted what was referred to as the "principle of logical sound usage."[25] It seems that for many writers, as soon as *some* sounds were clearly diegetic, *all* sound had to be so. As one reviewer of the part-talking boxer melodrama *Liebe im Ring* (*Love in the Ring*) (1929, dir. Reinhold Schünzel) complained: "Music has to be motivated in a film, as was the case in the scenes taking place in a variety theatre. But how did the music get to the fish market?"[26] Similarly, a reviewer of *Say It with Songs* (1929, dir. Lloyd Bacon) was fine with the music, as long as he/she could at least *imagine* that it came from a diegetic source, as in scenes set in a radio broadcasting studio, "but when the music continues plunking away at Jolson's home, staunch opposition arises within us."[27] Acclaimed critic Herbert Ihering called the little nondiegetic music that there was in Curtis (Kurt) Bernhardt's *Der Mann, der den Mord beging* (*The Man Who Committed the Murder*) (1930) a "fallacy," which evokes "confusion, not clarity."[28] One may wonder about the reasons for this sentiment after years of continuous musical accompaniment in the silent era, but as Michael Slowik has convincingly argued, the conceptual distinction between diegetic and nondiegetic sound was hardly relevant in silent film and became so only *with* the introduction of synchronized sound, which could now be clearly *marked* as diegetic (or not).[29]

The following two arguments partly derive from the logic argument but are nevertheless slightly different.

24 Quoted from O'Brien, *Movies, Songs, and Electric Sounds*, 79.

25 Seuffert, "Tonfilm-Impressionen," 2.

26 "Kritische Filmschau: Liebe im Ring," 10.

27 "Kritische Filmschau: Sag es mit Liedern," 7.

28 Ihering, "Der Mann der den Mord beging," 327.

29 Slowik, *After the Silents*, 55–61.

(3) The 'realism argument': *Nondiegetic music contradicts the inherent realism and objectivity of sound film.* For many writers, the novelty of sound film, especially of talkies, went hand in hand with a broader shift towards the 'real' and the 'objective' in the arts, which was often associated with the term "New Objectivity." The new technologies of sound reproduction, brought about by electricity, were seen as participating in this general shift, whereas the kind of live musical accompaniment known from the silent era now seemed to contradict this idea and was associated more with the world of fantasy and idealism. For several critics, the use of nondiegetic music in talking pictures created a 'clash' between the two concepts, leading them to take a firm stance against it, at least in the case of so-called "serious films."[30] For instance, Ernst Jäger, in his earlier writings in *Film-Kurier*, consistently advocated relinquishing the musical score. As he writes:

> Music *disturbs* during the strongest visual impressions. [...] The sound film demands the purification from music, for even the lively tango does not fit into the new and much truer scenic space of experience that the screen offers when the word is added to its shadows. [...] The serious sound film has put music back in its place.[31]

Similarly, film critic Kurt London rejected the musical score in newsreels because of their "atmosphere of objectivity," which he compared to that of an "illustrated newspaper."[32] The anonymous writers (or possibly only one) in *Deutsche Filmzeitung* also argued against nondiegetic music in talkies fairly consistently. The music in *Die Nacht gehört uns* reinfused "sentimentality" into an otherwise "objective" film;[33] the score in *Zwei Menschen* (*Two Souls*) (1930, dir. Erich Waschneck) meant a "breakaway from naturalism";[34] but the *lack* of a score in the German-language version of Greta Garbo's first talkie, *Anna Christie* (1930, dir. Jacques Feyder), fulfilled the "principle of sound film reality."[35] As early as March 1929, an article in *Deutsche Filmzeitung*

30 Obviously, there were other sound film genres that emphasized fantasy and dream-worlds, such as the operetta films.

31 Jäger, "Film-Kritik: Phantome des Glücks." Interestingly, this is an "incidental remark" in the middle of a review of a *synchronized* film without audible dialogue – *Phantome des Glücks* (*Phantoms of Happiness*) (1930, dir. Reinhold Schünzel) – in which Jäger envisions how the music *should* be treated *if* the film were a talkie.

32 London, "Wochenschau mit oder ohne Begleitmusik?"

33 "Kritische Filmschau: Die Nacht gehört uns," 5.

34 "Kritische Filmschau: Zwei Menschen," 4.

35 "Kritische Filmschau: Anna Christie," 4.

anticipates these arguments, commenting on Walter Ruttmann's first (and now presumably lost) short sound film experiment, *Deutscher Rundfunk* (*German Broadcast*) (1928):

> The mixture of reality (reproduction of sounds and noises without any artistic intentions) and ideality (musical accompaniment with claim to art) is wrong. You fall from one dimension of ideas into another [...] Any musical accompaniment as an addition in sound film has to fall behind in the race against sounds and noises.[36]

(4) 'The arbitrariness argument': *The use of nondiegetic music is arbitrary.* We can find several contemporary reviews complaining that the use of nondiegetic music betrayed a lack of artistic purpose and was simply used to 'fill in the blanks' that would have occurred at moments without speech, diegetic music, or background noises. It was a common sentiment among critics that music was often used as a "stopgap," as "sound plasters," or as a "hideous patchwork."[37] Obviously, this kind of usage contradicted the idea – advocated by many trade papers and intellectual writers – of films as purposefully created, well-conceived art works.

General opposition to nondiegetic music in 'serious' sound films became less frequent during the first half of 1931, just as the films released in the second half of the year – which contained more scoring – were in production. Thus, it seems that both filmmakers and critics gradually changed their stance and that a phase of fundamental and perhaps overly hasty reserve and suspicion was giving way to a period of habituation and fascination.[38] The pioneering efforts in films like *Boykott*, *Brand in der Oper*, and *Karamasoff* may have played a key role in convincing critics and filmmakers (and perhaps audiences at large) that nondiegetic music *could* be used in ways that were not confusing, illogical, arbitrary, or irreconcilable with realism but, in fact, quite appropriate for certain scenes, serving to intensify their overall meaning and atmosphere.

In this respect, German film critic Kurt London's book *Film Music*, published in British exile in 1936, is illuminating, as it retrospectively summarizes some of the sentiments from the transitional years. London draws a distinction between the musical score of the silent era and the sound era.

36 "Tonfilme," 7–10.

37 "Kritische Filmschau: Sonntag des Lebens," 4; "Kritische Filmschau: Flieger," 9; "Tonfilmkritik," 2. These reviews relate to *Sonntag des Lebens* (*The Devil's Holiday*) (1931, dir. Leo Mittler) and *Flight* (1929, dir. Frank Capra).

38 For a similar argument see Helmers, "The Transition from the Silent into the Sound Era," 136.

While musical accompaniment was a necessary *addition* to silent film but not really “a part of it,” the coming of synchronized sound meant that it became “an integral part of the sound-film” and thus demanded stronger justification: “In other words, music must have its meaning. There must be good reasons for its sound to be heard.”[39] While diegetic motivation was certainly an appropraite way to meet this demand (and London does bring in the ‘illogical argument’ at times), it seems that it wasn’t really the crucial point for him. On the contrary, London values the German operetta films and other sound films with long stretches of nondiegetic music. What he does oppose, however, is “background music” in the style of silent film’s “illustration.” The latter term had been widely used for silent film scoring and usually referred to the practice of assembling pre-composed music that fitted the general mood of specific scenes. However, since the music was not composed specifically for the film, it couldn’t help but be *indifferent*, to a degree, to the specific rhythm and development of a given scene. In sound film, as London and others felt, the score needed a ‘tighter connection’ to the images in some way. As he puts it: “Sound films need no illustration, but their music has to be the psychological advancement of the action. [...] It has to establish associations of ideas and carry on developments of thought.”[40] If we look at the initial scepticism towards scoring from this perspective, it seems likely that it was not nondiegetic music per se that was rejected but rather a specific way of using it – namely, a use that was still anchored in silent film’s “illustration” and that betrayed a lack of correspondence between music and images. Other, more satisfying uses had yet to be developed and *were* indeed being developed, increasingly so in the second half of 1931. In London’s words, as different ways of using nondiegetic music were explored, there were more and more “good reasons for its sound to be heard.”

Bibliography

Bauer, Alfred. *Deutscher Spielfilm-Almanach 1929–1950*. Berlin: Filmblätter Verlag, 1950.

Chaparral. “Dernier cri in Hollywood: Weg mit dem Sprechfilm.” *Film-Kurier*, March 12, 1929.

Cooke, Mervyn. *A History of Film Music*. Cambridge: Cambridge University Press, 2008.

39 London, *Film Music*, 126.

40 Ibid., 135.

Helmers, Maike. "The Transition from the Silent into the Sound Era: The Innovative Use of Sound in Pabst's Westfront 1918." *Music, Sound, and the Moving Image* 12, no. 2 (2018): 121–39.
Ihering, Herbert. "Der Mann, der den Mord beging." *Von Reinhardt bis Brecht: Vier Jahrzehnte Theater und Film, vol. III: 1930–1932*. Berlin: Aufbau-Verlag 1961.
Jacobs, Lea. "The Innovation of Re-Recording in the Hollywood Studios." *Film History* 24, no. 1 (2012): 5–34.
Jäger, Ernst. "Film Kritik: Der Mörder Dimitri Karamasoff." *Film-Kurier*, February 7, 1931.
Jäger, Ernst. "Film-Kritik: Phantome des Glücks." *Film-Kurier*, January 3, 1930.
Klaus, Ulrich J. *Deutsche Tonfilme*. Berlin: Klaus-Archiv, 1989.
"Kritische Filmschau: Anna Christie." *Deutsche Filmzeitung*, January 23, 1931.
"Kritische Filmschau: Der Schuss im Tonfilmatelier." *Deutsche Filmzeitung*, September 5, 1930.
"Kritische Filmschau: Die Königsloge." *Deutsche Filmzeitung*, July 11, 1930.
"Kritische Filmschau: Die Nacht gehört uns." *Deutsche Filmzeitung*, January 10, 1930.
"Kritische Filmschau: Flieger." *Deutsche Filmzeitung*, May 30, 1930.
"Kritische Filmschau: Liebe im Ring." *Deutsche Filmzeitung*, April 4, 1930.
"Kritische Filmschau: Sag es mit Liedern." *Deutsche Filmzeitung*, July 25, 1930.
"Kritische Filmschau: Sonntag des Lebens." *Deutsche Filmzeitung*, February 27, 1931.
"Kritische Filmschau: Zwei Menschen." *Deutsche Filmzeitung*, January 16, 1931.
London, Kurt. *Film Music: A Summary of the Characteristic Features of Its History, Aesthetics, Technique; and Possible Developments*. London: Faber & Faber, 1936.
London, Kurt. "Wochenschau mit oder ohne Begleitmusik?" *Der Film*, April 15, 1929.
Müller, Corinna. *Vom Stummfilm zum Tonfilm*. München: Fink, 2003.
O'Brien, Charles. *Movies, Songs, and Electric Sound: Transatlantic Trends*. Bloomington: Indiana University Press, 2019.
Schacht, Roland. "Tonfilmdramaturgie." *Film-Kurier*, April 22, 1929.
Seuffert, L. v. "Tonfilm-Impressionen." *Deutsche Filmzeitung*, January 30, 1930.
Slowik, Michael. *After the Silents: Hollywood Film Music in the Early Sound Era, 1926–1934*. New York: Columbia University Press, 2014.
Thiele, Wilhelm. "Operette im Tonfilm." *Film-Kurier*, February 1, 1930.
"Tonfilme." *Deutsche Filmzeitung*, March 8, 1929.
"Tonfilmkritik. Äußerungen in der 'Deutschen Presse." *Deutsche Filmzeitung*, June 13, 1930.
Wedel, Michael. *Der deutsche Musikfilm: Archäologie eines Genres 1914–1945*. München: text + kritik, 2007.
Wiegand, Daniel. "'Islands of Sound in the Silent Flow of Films': German Part-Talkies around 1930 as a Hybrid Medium." *Historical Journal of Film, Radio, and Television Historical Journal of Film, Radio, and Television* 24, no. 3 (2022): 427–50. https://doi.org/10.1080/01439685.2021.1976916

About the Author

Daniel Wiegand is Assistant Professor of Film Studies at University of Zurich. He is the author of *Gebannte Bewegung. Tableaux vivants und früher Film in der Kultur der Moderne* as well as co-editor of *Film Bild Kunst. Visuelle Ästhetik des vorklassischen Stummfilms* (edited with Jörg Schweinitz, 2016). Recent articles include "'Islands of Sound in the Silent Flow of Film': German Part-Talkies around 1930 as a Hybrid Medium," in *Historical Journal of Film, Radio and Television* (2022).

14. The Sounds of War: Reflections on WWI Films of 1930

Martin Holtz

Abstract: This contribution considers how four war films released in 1930 use sound: the trench warfare dramas *All Quiet on the Western Front* and *Westfront 1918*, the air warfare film *The Dawn Patrol*, and the comedy *Doughboys*. These films present the sounds of war as a tapestry of noise to create a feeling of disablement as the essential characteristic of World War I, thereby capturing the entropic and fragmented nature of modern society reeling from economic collapse amidst technological progress. In this way, they provide a self-reflexive counter-discourse to the role of the sound film in positing against its narrative of progressive and utopian "electrical entertainment" a critique of film sound as an emblem for a machine age hostile to humanity.

Keywords: war films, noise in early sound film, human voice, silencing, modernity

According to most scholars, by 1930 sound was "consolidated into [an] unostentatious presence"[1] in American films. After a brief period of foregrounding the ability of the medium to reproduce synchronized sound as a special effect, removed from direct narrative purposes, sound had entered a stage of "containment,"[2] modulated and orchestrated so as to provide service to narrative and spatial coherence and continuity without drawing attention to itself as a technological phenomenon.[3]

1 Crafton, *The Talkies*, 355.

2 Ibid., 16.

3 Altman, "Sound Technology," 46–47; Bordwell/Staiger/Thompson, *Classical Hollywood Cinema*, 308; Crafton, *The Talkies*, 355–80; Lastra, *Sound Technology and American Cinema*, 180–215; O'Brien, *Cinema's Conversion to Sound*, 44–63.

Wiegand, D. (ed.), *Aesthetics of Early Sound Film: Media Change around 1930*.
Taylor & Francis Group, 2023
DOI 10.5117/9789463727372_CH14

While I don't dispute this general narrative of the development of sound film in American and most Western cinemas, I want to point out the peculiar role of the war film genre in this respect, more particularly the string of films released in 1930 that use World War I as a backdrop: *All Quiet on the Western Front* (dir. Lewis Milestone), *Westfront 1918: Vier von der Infanterie* (*Westfront 1918*, dir. G. W. Pabst), *Doughboys* (dir. Edward Sedgwick), and *The Dawn Patrol* (dir. Howard Hawks). Donald Crafton argues that the development of the sound film in the 1920s was "plugged" into a culture-wide narrative of the progression of science and technology heralding "the promise of a better future,"[4] which is why early sound films often staged moments of synchronized sound as an "encounter […] with technological modernity."[5] If by 1930 the cinema had overcome this staging of synchronized sound as a special effect, then the war film returns to the treatment of sound as an encounter with technological modernity, only not in the form of benevolent scientific advancement, but as traumatizing shock. Sound becomes, despite its narrative integration and modulation, a foregrounded, disruptive, self-reflexive vehicle criticizing the progress that it signifies.

Particularly the trench warfare films use the sounds of machine guns and grenade explosions to create a tapestry of noise that transcends the characters' and audience's capacity to fully register and decode its informational content. The sounds create an off-screen space of aural engulfment that disorients and hence incapacitates characters and viewers alike, correspondent with the overriding sensation of attrition as the essential characteristic of World War I.[6] In this way, sound in the war films is used emblematically to signify the silencing of the human individual in a deafening, destructive machine age. The films resonate with a contemporary audience in communicating the entropic, fragmented, and coercive existence in a modern(ist) (urban) society amidst technological progress and leaving

4 Crafton, *The Talkies*, 21.

5 Lastra, *Sound Technology and American Cinema*, 5.

6 A number of scholars have pointed out how World War I decisively shifted the experience of war at the front towards aurality. Because of the limited use of sight in the trenches, hearing became the essential sense to guide the soldiers on the battlefields. The ear was disciplined, drilled to perform active duties of gathering and interpreting information amongst the sounds of battle in order to facilitate survival. At the same time, the discourse of disciplined and controlled listening was more often than not undermined by the reality of trauma, as the ear was simply not able to filter and process the incessant barrage of noise (Encke, *Augenblicke der Gefahr*, 152–93; Volmar, "In Storms of Steel," 228–32; Müller, "Akustik des Krieges," 91). The early sound films convey precisely this sensation of aural "overwhelm" and run counter to the narrative of disciplined listening.

the individual incapacitated and overwhelmed by economic, social, and political discourses beyond their control.[7]

All Quiet on the Western Front

All Quiet on the Western Front has long been recognized as a groundbreaking example of the genre. In critical discourse it is routinely stressed how the film established genre conventions in terms of structure, aesthetics, character constellation and development, and tonal ambivalence.[8] The addition of sound was effectively used in the marketing of the film and also emphasized in subsequent scholarship as boosting the film's realism and innovation.[9]

Already the title of the film (and of the novel by Erich Maria Remarque that it is based on) point to the significance of sound. Its irony is that the most significant event in the story, the death of the protagonist via the loud noise of a sniper's rifle, happens on the day the official report suggests quietude.[10] What the title and ending communicate is the mismatch between a national, political officialdom and the experience of the individual who is silenced by the noise of a bullet as much as by the authority of the official report. The use of sound in the film reflects this silencing of the individual by the forces of society at large embodied by a war machine of deafening noise.

The opening sequence introduces this theme and its aesthetic implementation. It starts with a parade of marching soldiers amidst loud, rhythmic music and the cheers of the public lining the streets. When the camera moves from the streets through incongruously open windows into a classroom, the noise of the mass's fervor drowns out the voice of the teacher, imposing itself onto the individual.[11] Once the soundtrack gives prominence to the teacher's voice, it becomes clear that he is a spokesperson for the glory of

7 Geisler, "Battleground of Modernity," 99–102; Eksteins, "All Quiet on the Western Front," 31. In fact, the discourses surrounding the production of noise as a health crisis in the interwar years convey the centrality of noise as a manifestation of technological coercion (Mansell, "Neurasthenia, Civilization, and Sounds of Modern Life," 280).

8 Williams, "Film and Mechanization of Time," 186; Geisler, "Battleground of Modernity," 91; Winkle, "Phantasmagorien des Krieges," 370.

9 Eksteins, "All Quiet on the Western Front," 29; Kelly, "In the Trenches, on the Screen," 14, 16.

10 In this context the English translation of the German book title *Im Westen nichts Neues* has indeed a more meaningful significance.

11 Historically, sound became a decisive factor in what Daniel Morat calls the "collective self-mobilization" ("Cheers, Songs, and Marching Sounds," 178) of the German population, which occurred in precisely such sound-filled gatherings where collectivity and fervor were rehearsed via the rhythmic and enervating spectacles of "cheers, songs, and marching" (177).

sacrificing your life for the country in war. The sound dissolve suggests a back translation of the unintelligible noise of fervor into the sanctioned propaganda speech of the state. Crucially, for most of the speech we see close-ups of the students' faces, suggesting a determining superimposition of the state's discourses onto the minds of the individuals. To use Michel Chion's terminology, the teacher's voice becomes "acousmatic," "sound that one hears without seeing its source,"[12] to which Béla Balázs adds: "If the sound or voice is not tied up with a picture of its source, it may grow beyond the dimensions of the latter. Then it is no longer the voice or sound of some chance thing, but appears as a pronouncement of universal validity."[13] The film illustrates this elevation of the teacher's speech from individual expression to universal dictum by which the students are interpellated (to use Althusser's term) with the state's ideology, which animates them to go to war.

Off-screen sound also characterizes the experience of war. We are introduced to the battle space by the sounds of whistling bombs and explosions when the soldiers arrive by train. The whistling noise illustrates war's disruption of spatial coherence and causality because it indicates the presence of a threat without localizing its causal source, resulting in an effect of destruction that erupts in the frame without allowing for a proper orientation to react accordingly. Silent film had to show the source of threat in order for it to exist, sound film suggests its continuous presence while making it ungraspable and so adds the crucial elements of randomness and confusion to the experience of war.

The incapacitating effects of sound are nowhere better illustrated than in the scene where the soldiers have to hold out in a shelter through a continuous bombardment by the enemy. While we hear explosions going off, we see the faces of the terrified young and inexperienced soldiers constantly expecting a lethal impact. Once again off-screen sound imposes itself on the faces of the individuals, this time rendering them immobile and powerless. The "universal validity" of the bomb noise becomes the aural manifestation of their precarious existences under the threat of random annihilation. The suicidal escape from the shelter to meet the noise head on is the desperate attempt to regain some sort of control over oneself by at least determining the moment of death.

There is a character in the film who offers a survival strategy within all this chaos. Katczinsky (Louis Wolheim) is portrayed as a seasoned veteran

12 Chion, *Film: A Sound Art*, 39.

13 Balázs, "Theory of the Film: Sound," 120.

who becomes a mentor figure to the young recruits, particularly to the protagonist Paul Bäumer (Lew Ayres). His ability to survive and maintain some sort of control over his existence amidst hostile conditions is explicitly linked to his competence of reading noise correctly. He is the ideal of the "trained listener"[14] as he can decipher the whistles of the incoming bombs and react accordingly. When everyone around him ducks in terror, he stands unmoved, not affected by the terror of sound. Hence, Kat suggests the possibility that sound need not be the manifestation of random destruction but rather the signal by which to navigate through a hostile machine age.

This ability is put to the test in the actual combat scenes of the film. These sequences were shot silent with sound added in post-production, much like earlier synchronized films had done. A tapestry of noise consisting of whistles, explosions, gunshots, yells and screams are laid over the shots, sometimes matching the image in moments of synchronicity but ultimately adding many more battle sounds to what can be seen. The general effect of this approach is what Siegfried Kracauer called "exploding the borders of the picture."[15] The sounds suggest action that transcends the frames of what the camera can show. It is action beyond the image, beyond perception, and hence beyond the individual's ability to wholly grasp and comprehend it. Visually the film adds to the sense of an exploding border of the frame with its panning dolly shots: in one instance where the camera moves over a trench looking down while attacking soldiers are pouring into it to engage in a melee fight, and, most famously, in the machine gun scenes, where the camera moves along rows of incoming soldiers as they are mowed down by gunfire, which is intercut with static shots of the machine gun "panning" effortlessly from one side to the other, whereby machine gun and camera, both products of the machine age, are eerily fused.[16] While this is a predominantly visual moment, the rhythmic efficacy of the killing machine acquires an unforgiving persistence when we actually hear it. The effectiveness of the machine is contrasted with the ineffectiveness of the (human) effort. The French attack is warded off, the counter-attack fizzles out, the fight ends in a stalemate. In this way, the chaos conveyed by image and sound is joined by an air of futility, suggesting a meta-commentary on the absurdity of the display of technological spectacle at the expanse of story or, by metaphorical extension, meaningful human progress.

14 Encke, *Augenblicke der Gefahr*, 123, 182–85 (my translation).

15 Quoted in Fisher, "Landscapes of Death," 273.

16 Williams, "Film and the Mechanization of Time," 187; Kelly, "In the Trenches, on the Screen," 16–17.

This brings me to the representation of the protagonist. The discourses that impose themselves on Paul on an aural level, the population's fervor, the teacher's speech, the sounds of war, are contrasted with the ineffectuality of his voice. A crucial scene occurs when he gets trapped in a bomb crater during an attack and stabs a French soldier falling into it, who then bleeds to death. In the course of the night, Paul develops remorse over his deed and pleads with the dead man for forgiveness. We hear his pleading off screen while we see the immovable face of the dead man. In this case, the voice does not attain a universal validity, as in the classroom scene before, rather, the ineffectuality of his attempts to relieve his guilt over his actions, is illustrated by the lack of any reaction from his addressee. The individual fails to impose himself on his environment.

When Paul is in a hospital, he fears that he will be taken away to die as a hopeless case. His pleadings are disembodied, as doctors and nurses crowd around him so that he disappears from the view of the camera. The medical personnel not only ignore Paul's desperate utterings, it murmurs amongst itself as it imposes its own narrative onto the helpless patient, whose voice is eventually drowned out. Even if Paul will eventually come out alive from the hospital, the scene illustrates his utter lack of control over his survival.

When Paul comes back home and visits his old school, he finds the teacher indoctrinating another class with speeches of war's glory. This time, Paul tries to reclaim his narrative, telling the class about the reality of inglorious death and desperation on the battlefield. The camera grants him a medium shot during his speech, even if it is framed by the teacher in the foreground. Reaction shots of the students illustrate their resentment of the counter-narrative rather than any correction of their romantic ideas about war. The scene communicates to the viewer the veracity and importance of Paul's narrative and voice while at the same time staging its ineffectiveness in its diegetic context.

Paul returns to the front where he reunites with Katczynski. Kat's death serves as the ironic counterpoint to the scene in which he demonstrated his ability to read the sounds of war correctly. Hit by an incoming shell in the leg, he is carried by Paul to the nearest field hospital when another shell hits nearby. This one delivers a mortal blow to his neck, which remains unnoticed by Paul who proceeds to talk jocularly to another dead man. The interpreter of the sounds of war dies voicelessly, killed by a bomb he is unable to read properly while the less capable soldier escapes unscathed, which calls into question Katczynski's competence in the first place. And Paul's voice remains just as ineffectual as his efforts to rescue his friend. Everything he does is rendered futile.

Paul's death scene delivers the final irony. While the noise in the combat scenes is characterized by permanence and disorientation, Paul's death is portrayed with clear causality. As he reaches out over the edge of the trench to touch a butterfly, a reverse shot reveals a French sniper taking aim. The source of threat is thus clearly established as the soundtrack remains ominously quiet. When we hear the gunshot, we don't see the rifle but Paul's jerking hand before it stops moving. It is the *sound* of technological effectiveness that has killed the man and that imposes itself onto the last twitches of humanity reaching out for life.

Westfront 1918

The German production *Westfront 1918: Vier von der Infanterie* is often discussed in relation to *All Quiet on the Western Front*, emphasizing similarities in their anti-war stance despite different cultural contexts and a similar approach to sound despite differences in their cinematography and editing.[17] The opening of the film illustrates the use of off-screen sound as a similar discourse of universal validity imposing itself on powerless individuals. We see a group of soldiers frolicking in a tavern engaged in a playful tryst over a woman when one soldier slaps the behind of another, but instead of a slap we hear an explosion rattling the inn. On the one hand, the explosion functions as a violent imposition of war on the heretofore ludic activities of the soldiers, drowning out their joy and instantly changing their behaviour, as they anxiously turn to expect their detachment to the front. On the other hand, the explosion can also be seen as the logical extrapolation of the playful violence that the soldiers engage in, as a comment on the limitless violence that can spiral out from human foibles in an unpredictable and uncontrollable fashion, exploding the borders of the frame. After all, especially World War I illustrates how every war has small beginnings.[18] A subsequent shot shows an outside view of the tavern, the space of refuge and human joy, overlaid with the yells of "Still gestanden!" ("Stand to attention!") as the soldiers are made to march to the front off-screen, the mismatch of sound and image illustrating the yearning for a communal sanctuary undermined by the war.[19]

17 McGuire, "Filtering and Interpreting the Great War," 667, 675; Geisler, "Battleground of Modernity," 91; Winkle, "Phantasmagorien des Krieges," 370.

18 Fisher, "Landscapes of Death," 273.

19 Ibid., 275.

As Jaimey Fisher points out, *Westfront 1918* is an anti-mobilization film expressing the anti-war spirit of large portions of Weimar Germany and as such emphasizes the immobilizing effects of war.[20] Human voices at the front are routinely presented in a whispering tone, the vocal restraint emblematic of a greater sense of constriction in this hostile environment.[21] As bombs start to hit, the soldiers in the trenches strain to locate their origins when one of them realizes that some emanate from cannons behind their lines. The confusion about source and causality inherent in off-screen sound is used by the film to illustrate how the individual, caught in a no-man's land between friendly and enemy fire, is churned up by discourses beyond his control. The feeling of victimization by random violence finds its most extreme manifestation when some soldiers are buried underneath a collapsed shelter and comrades struggle to dig them out. Human sounds are reduced to strained breathing, frantic yells, and ultimately silence, as war is rendered aurally as a suffocating experience.

In a later scene we hear a wounded soldier crying for help from no-man's land while we see the desperate faces of the soldiers in the trenches. The film remains ambivalent about whether the wounded soldier is indeed the missing comrade who got lost in a skirmish days earlier or one of the enemy. Nationality, the films suggests, does not and should not guide compassion.[22] The off-screen yells communicate a twofold powerlessness of the individual, on the one hand signaling the helplessness of the wounded soldier, whose voice lacks the agency to mobilize assistance, on the other hand imposing themselves onto his comrades who are similarly incapacitated in venturing out into hostile territory to help him.[23]

The combat scenes in the film are less mobile and editing-driven than in *All Quiet on the Western Front*. They tend to feature static long shots with various planes of action where actions erupt into the frame, fill it, and then leave it again, by which the limited capacity of vision is emphasized.[24] The image is either over-filled, overwhelming perception, or empty, prompting a lack of perception. The soundtrack works in similar ways, mixing on-screen and off-screen explosions and whistles and stretches of eerie silence in between, whereby it adds a sense of disorientation.

20 Ibid., 269; Fisher, "Haptic Horrors of War," 58.

21 Fisher, "Haptic Horrors of War," 61.

22 In this way the film follows the same sentiment as G. W. Pabst's subsequent companion film *Kameradschaft* (1931) in which French and German miners collaborate to free their pitmen from a collapsed mine.

23 Fisher, "Haptic Horrors of War," 65.

24 Winkle, "Phantasmagorien des Krieges," 371–72.

Figure 14.1: The Lieutenant (Claus Clausen) screams madly in *Westfront 1918* (1930, dir. G. W. Pabst).

The battle noises culminate in a contrast with the ineffectual human voice. As the Lieutenant (Claus Clausen) sees his men charging into another attack that will brutally decimate them, he incongruously shouts "Hurrah, Hurrah" while standing on a heap of corpses. The voice expresses a delusional state, a mismatch with the image of universal defeat. It stands as a manifestation of not only limited but misguided perception of the individual, whose shouts of glory as propagated by the state are exposed as lunatic ravings of a maniac (Fig. 14.1).

The two survivors from the group of protagonists are brought into a hospital where the noise of groans, screams, and frantic medical personnel overlaid by a continuing barrage of off-screen explosions serves as the final discursive imposition onto the voices of the dying soldiers, who have completely retreated into an internal world of dreams and memories, muttering barely audible words until death constricts their voices permanently. The explosions even continue over the end credits, a tentative "Ende?", suggesting their potential permanence and ultimate triumph over human life in a frightening expression of "universal validity."

Doughboys

Even if *Doughboys* is a comedy tonally completely different from the downbeat *All Quiet* and *Westfront*, it shares with those films similar concerns of showing the individual churned up by war and using sound as the expressive device to communicate his powerlessness. The persona that Buster Keaton had developed over the course of his silent films, most prominently in *The General* (1926, dir. Clyde Bruckman, Buster Keaton) with its Civil War scenario, is an appropriate character here as he routinely encounters adverse circumstances with alternations of incredulous luck, (self-)surprising competence, and circumstantial heroism, thereby highlighting how much the individual is at the mercy of random forces beyond his control.

Keaton's character Elmer is an industrial heir of leisure spending his time romancing a factory girl. However, his (unsuccessful) dating efforts are overlaid by a parade of soldiers and the "Extra! Extra!" yells of the newsies announcing America's entry into the war. The militaristic discourse aurally imposes itself onto Elmer's romantic desires (the irony of course being that his heroic exploits in war will win him the girl). When his chauffeur joins the military, Elmer accidentally ends up in a recruitment office where he tries to find a new chauffeur when he, while swooning for his crush and imagining marriage vows, ends up being recruited himself. The recruiting secretary, representative of the militaristic discourse, thus imposes a (mis)interpretation onto Elmer's words that determines his further existence.

In boot camp, the military discourse imposes itself on musical interludes and soft-spoken recruits in the form of blaring trumpets and the sergeant's yells, probably nowhere more illustratively than in the scene where the recruits are supposed to emulate the sergeant's bloodcurdling war cry, only to manage a gentle "woof." The joke is in the failure of the recruits to live up to the savage volume of war. Noise is a nuisance, so why participate in it, the film seems to say in an eerily appropriate comment on Keaton's own sound film career.

The film then takes the ironies of *All Quiet on the Western Front*'s title into comedic territory when it contrasts the official reports of "sunny France" and "quiet sector" with the reality of torrential rain and seismic explosions. The mismatch between report and reality, expressed in straight (sound) cuts between speechifying officers and noisy panoramas comedically illustrates the duplicity of the discourse while highlighting its victimization of the individual who after all has to endure everything it submits him to.

What characterizes the combat scenes of *Doughboys* is the inability of the individual to master the sounds of war, which appear to have their own agency. When Elmer ventures out into no-man's land with a gas mask, the noise of his breathing is amplified on the soundtrack, prompting him to fear his detection by the enemy, as the machines of war turn the most rudimentary and vital body function into a source of threat. There are several instances when the accidental production of noise or the absence of an expected noise, e.g., when an explosion goes off in a ditch, a bomb turns out to be a dud, or Elmer is startled awake and unintentionally fires his rifle, set off a chain of consequences that lead to decisive battle events. Noise is embodying the contingent and accidental in war, illustrating a mismatch between intention and outcome of actions, as Elmer is turned into an unlikely hero who saves and gets the girl while intermittently fraternizing with and ultimately vanquishing the enemy. The absurdity of random causality is emphasized by the kinds of noises we hear, which are peculiarly lacking in fidelity. Bomb whistles and explosions sound like cheap joke article sound effects. This also highlights the incongruous artifice of the noise, by which the film lampoons the presumptions of sound film to convey realism.[25]

Appropriately, the film ends with a short post-war scene in which Elmer presides over an officious corporate meeting when everyone is startled by what sounds like machine gun fire and ducks for cover with the former drill sergeant, who now works as a cleaner for Elmer, immediately reverting back to his soldierly habitus of yelling orders. Only *we* can see that it is a jackhammer which produced the noise. Thus, sound retains its war-like quality even beyond the battlefield in the modern, machine-dominated society and also its potential for misinterpretation and causing unintended consequences.

The Dawn Patrol

An examination of American war films of 1930 would not be complete without a look at aviation films. Inspired by the success of the synchronized silent film *Wings* (dir. William A. Wellman) three years before, these films are often seen as a chivalrous adventure counterpart to the grim infantry films, glorifying war as modern knightly romances and showcasing the capabilities of the medium and modern technology with breathtaking

25 Müller, "Akustik des Krieges," 105.

action sequences.[26] While this view is certainly applicable to *Wings* and 1930's *Hell's Angels* (dir. Howard Hughes), *The Dawn Patrol* has a distinctly different flavor, sharing with the infantry films a disillusioned take on war by utilizing sound to illustrate the limited agency of the individual.

The film opens in medias res with the chaos of an air battle. Where dogfights in *Wings* and *Hell's Angels* are given polished continuity editing, explanatory dialogue or intertitles, and, for *Hell's Angels*, even a scale-matched soundtrack, the approach in *The Dawn Patrol* is decidedly messy. The editing does little to isolate the parties, establish causality, or match the blaring cacophony of engine noise and gun fire with the images. Back projection is extensively used as opposed to the daring stunt work in *Wings* and *Hell's Angels*. Close-ups of the confused faces of pilots and a crash and burn outcome suggest that war is not a pretty sight (and sound) but a chaotic mess. Subsequent montage sequences employ layered imagery and sounds to deepen the idea of impenetrable messiness and confusion. This sense of limited agency and attractiveness of war is important for the film's overall conflict, which revolves, in true Hawksian fashion, around professionalism, the dispassionate commitment to a job vs. a resignation into cynicism about its meaninglessness.[27] The question the film asks is how one can maintain a professional attitude engaging in something that is so resistant to dignity and agency.

The commanding officer Major Brand (Neil Hamilton) represents cynical despair as he is forced by the higher ups to send his men into suicidal missions without having the moral compensation of participating himself. Sitting at his desk, he is expressing his furious discontent over orders into a telephone, but we never hear any sound emanating from it. The machine is silent (Fig. 14.2). The absence of sound signifies the deliberate ignorance towards human concerns of the high command as much as of mechanized modernity. The human voice is rendered impotent, unworthy of response.

This is not the only instance in the film in which humans are reduced to having to listen to machines. Several scenes show the CO awaiting the return of his men from battle by listening to the sound of the planes as they land. The absence of sound signifies death, the pilot who has not returned, so that in both cases, telephone and returning planes, the stoic silence of machines renders the individual quasi or factually dead. There are also moments when the sounds of incoming planes are misinterpreted, generating false hopes,

26 McGuire, "Filtering and Interpreting The Great War," 670; Kelly, "In the Trenches, on the Screen," 12.

27 Wood, *Howard Hawks*, vii.

Figure 14.2: Major Brand (Neil Hamilton) is listening to a silent telephone in *The Dawn Patrol* (1930, dir. Howard Hawks).

as when an enemy plane approaches only to drop a message about a shot down fighter. The film thus presents two extremes of sound as signifiers of man's lack of agency in war: the chaotic, layered noise of the air battles and the impotent shouts of anger from the CO vs. the stoic silence of the telephone and the death-heralding absence of engine noise.

If loudness and silence represent the conditions and expressions of powerlessness that lead to cynicism and frustration, then the measured, controlled use of sound expresses professionalism. The protagonist of the film, ace pilot Dick Courtney (Richard Barthelmess), is characterized in such a way. His reaction to the furiously expressed orders from Brand is a curt "Right." The stoic affirmation bespeaks a sense of efficient communication, the absence of noise in favour of signal. This efficiency is also transported in the ritualistic speech occasions that Courtney engages in: the staff reports after battles and the starting procedures before every battle, all characterized by the unagitated use of speech to transmit essential information. Probably the most poignant ritual Courtney engages in is the communal singing in a bar after a battle. Significantly, also a shot down enemy pilot is admitted into the temporary reverie, illustrating how the harmonious approach to sound in song generates a peaceable coexistence and in fact the kind of chivalry that war appears to wipe out.

The mastery of sound that Courtney displays on the ground also characterizes his performance in battle. Bomb drops and machine gun attacks marry image and sound in a way that signifies clarity of purpose and efficiency. The tragedy of the film is that Courtney's order-imposing performance in battle is curtailed when he is appointed CO and, ironically because of his success as a fighter, forced into a position of powerlessness. When he has to experience the death of his men, among them an idealistic but inexperienced pilot, from a position of passivity, he overcompensates by acting heroically, going on a suicidal bombing mission alone and against orders. This is as much an assertion of agency as it is an abandonment of professionalism, a conundrum forced upon him by the military high command. He manages to destroy the munitions factory of the enemy, displaying the daring mastery of noise and chaos as in prior battles, but is shot down and killed by enemy fighters in his attempt to return. The ending is ambivalent. Business at the base continues in orderly fashion with Courtney's protégé taking over command, but within a structure that the whole film has exposed as detrimental to agency. And yet, the fact that this continuity is expressed with measured acceptance, a curt "goodbye" after the message of Courtney's death is dropped from an enemy plane, and a similarly curt affirmation of succession, suggests a triumph of professionalism in the mastery of sound. Hence, the film uses sound to characterize war and military structure as resistant to order and agency by alternately presenting it as noise or silence. At the same time, it champions the precarious value of the individual's professionalism, which asserts order and control in its measured communication of signal and expression of harmony.

Conclusion

By 1930 sound film was well established in the USA. The focus on sync-sound as spectacle had worn off, and film practice sought to integrate and subordinate the soundtrack to transparent narrative purposes. War films follow that trend in featuring modulated soundtracks that combine direct sound with post-synchronized effects, but they differ in foregrounding sound as spectacle. More specifically, the sounds of war are presented meaningfully as noise that signifies the meaninglessness of war, which is achieved by producing an aural environment that destabilizes causality. The amount of aural information (in the form of explosions, gunfire, bomb whistles) transcends the visual capacity of the image and the individual's ability to register and interpret it. The result is the presence of random and permanent threat, which negates the performance of meaningful actions. The individual is at the mercy of forces beyond his control, which is reflected in the films' use of voice. If human voices are not drowned out by the noise of the battlefield, they are suffocated, ignored, or rendered ineffective and meaningless. With this approach, the films critically comment on the status of humanity in a modern age characterized by technological advancement and a felt social neglect and they provide a self-reflexive counter-discourse to dominant conceptions of sound film in positing against its narrative of progressive and utopian "electrical entertainment"[28] a critique of film sound as an emblem for a machine age hostile to humanity.

Bibliography

Altman, Rick. "The Evolution of Sound Technology." In *Film Sound: Theory and Practice,* edited by Elisabeth Weis and John Belton, 44–53. New York: Columbia University Press, 1985.

Balázs, Béla. "Theory of the Film: Sound." In *Film Sound: Theory and Practice,* edited by Elisabeth Weis and John Belton, 116–25. New York: Columbia University Press, 1985.

Bordwell, David, Janet Staiger and Kristin Thompson. *The Classical Hollywood Cinema: Film Style & Mode of Production to 1960.* New York: Columbia University Press, 1985.

Chion, Michel. *Film, A Sound Art.* Translated by Claudia Gorbman. New York: Columbia University Press, 2009.

Crafton, Donald. *The Talkies: American Cinema's Transition to Sound, 1926–1931.* Berkeley, Los Angeles, London: California University Press, 1997.

Eksteins, Modris. "All Quiet on the Western Front." *History Today* 45, no. 11 (1995): 29–34.

Encke, Julia. *Augenblicke der Gefahr. Der Krieg und die Sinne, 1914–1934.* München: Fink, 2006.

Fisher, Jaimey. "The Haptic Horrors of War: Towards a Phenomenology of Affect and Emotion in the War Genre in Germany, 1910s to 1950s." *Seminar* 50, no. 1 (2014): 51–68.

Fisher, Jaimey. "Landscapes of Death: Space and the Mobilization Genre in G.W. Pabst's *Westfront 1918* (1930)." In *The Many Faces of Weimar Cinema: Rediscovering*

28 Crafton, *The Talkies*, 21.

Germany's Filmic Legacy, edited by Christian Rogowski, 268–85. Rochester: Camden House, 2010.

Geisler, Michael. "The Battleground of Modernity: *Westfront 1918* (1930)." In *The Films of G.W. Pabst: An Extraterritorial Cinema*, edited by Eric Rentschler, 91–102. New Brunswick, NJ: Rutgers University Press, 1990.

Kelly, Andrew. "In the Trenches, on the Screen: World War I on Film." In *The Wiley-Blackwell History of American Film Vol. 1: Origins to 1928*, edited by Cynthia Lucia, Roy Grundmann and Art Simon, 1–16. Hoboken: Wiley-Blackwell, 2011.

Lastra, James. *Sound Technology and the American Cinema: Perception, Representation, Modernity*. New York: Columbia University Press, 2000.

Mansell, James G. "Neurasthenia, Civilization, and the Sounds of Modern Life: Narratives of Nervous Illness in the Interwar Campaign against Noise." In *Sounds of Modern History: Auditory Cultures in 19th- and 20th-Century Europe*, edited by Daniel Morat, 278–302. New York: Berghahn, 2014.

McGuire, John Thomas. "Filtering and Interpreting The Great War: *All Quiet on the Western Front, Journey's End, Westfront 1918*, and Their Perspectives on World War I." *Quarterly Review of Film and Video* 33, no. 7 (2016): 667–80.

Morat, Daniel. "Cheers, Songs, and Marching Sounds: Acoustic Mobilization and Collective Affects at the Beginning of World War I." In *Sounds of Modern History: Auditory Cultures in 19th- and 20th-Century Europe*, edited by Daniel Morat, 177–200. New York: Berghahn, 2014.

Müller, Corinna. "Akustik des Krieges. Der Erste Weltkrieg als akustisches Ereignis im frühen Tonfilm." In *Der erste Weltkrieg im Film*, edited by Rainer Rother, Karin Herbst-Messlinger, 90–111. München: *Der erste Weltkrieg im Film*, 2009.

O'Brien, Charles. *Cinema's Conversion to Sound: Technology and Film Style in France and the U.S.* Bloomington: Indiana University Press, 2005.

Volmar, Axel. "In Storms of Steel: The Soundscape of World War I and Its Impact on Auditory Media Culture during the Weimar Period." In *Sounds of Modern History: Auditory Cultures in 19th- and 20th-Century Europe*, edited by Daniel Morat, 227–55. New York: Berghahn, 2014.

Williams, David. "Film and the Mechanization of Time in the Myth of the Great War Canon." *English Studies in Canada* 41, no. 2–3 (2015): 165–90.

Winkle, Ralph. "Phantasmagorien des Krieges – Kontinuität und Diskontinuität ästhetischer Muster in Filmen der dreißiger und vierziger Jahre." *Krieg und Literatur* 12 (2006): 369–81.

Wood, Robin. *Howard Hawks*. New ed. Detroit: Wayne State University Press, 2006.

About the Author

Martin Holtz is Assistant Professor in American Studies at University of Graz. He is the author of *American Cinema in Transition: The Western in New Hollywood and Hollywood Now* (2011) and *Constructions of Agency in American Literature on the War of Independence: War as Action, 1775–1860* (2019). Recent articles include "The Female Avenger in Post-9/11 Westerns" (2020) and "Machines in the Garden: Technology and the Western in the 1960s and 70s" (2018).

15. Urban Noise in the Early Italian Sound Film *Gli uomini, che mascalzoni ...*

Nadine Soraya Vafi

Abstract: Using *Gli uomini, che mascalzoni ...* (*What Scoundrels Men Are!*) (1932, dir. Mario Camerini) as a case study, this chapter argues that urban noise in early Italian sound film functions as a sonic reflection of Italian modernity and modernization under fascism: cinematic noises of cars and traffic resonated with contemporary developments in urban planning, and the almost documentary portrayal of Milan and its soundscape became part of a realist aesthetic that acknowledged working-class everyday life and concerns, anticipating 1940s Italian neorealism. Furthermore, Camerini's use of the soundtrack betrays a tendency towards experimentation enabled by a liberal climate of artistic exchange, which was possible despite fascism.

Keywords: noise, urbanism, modernity, fascism, working class

The various sound effects and noises heard in Italy's earliest sound films were made possible by the new recording and synchronizing technologies. They allowed, most importantly, for another layer to be woven into the foregrounded sounds of dialogue and music. Much like in other countries during the conversion to sound film, these noises and their narrative and aesthetic uses sparked vivid discussions among Italian film theorists and filmmakers (such as Sebastiano Arturo Luciani, Anton Giulio Bragaglia, or Luigi Pirandello), who debated whether sound did, in fact, enhance or rather diminish the expressive art of film.[1]

1 Bertellini, "Dubbing *L'Arte Muta*," 43–59; Sisto, *Film Sound in Italy*; Pitassio, "Technophobia and Italian Film Theory in the Interwar Period"; Dotto, "How We Learned to Listen to Noise."

Wiegand, D. (ed.), *Aesthetics of Early Sound Film: Media Change around 1930*.
Taylor & Francis Group, 2023
DOI 10.5117/9789463727372_CH15

Defining noise is a rather complex issue.[2] In this chapter, I will speak of noise not as negatively defined, unwanted sound (such as the inevitable ground noise in early sound films) but as intentionally created diegetic sounds, predominantly industrial, mechanical, man-made sounds, or crowd noise, which reflect the urban environment. This metropolitan noise was an unmissable part of the "multiple modernities"[3] across the globe and included the sounds resulting from technological inventions and infrastructural shifts in urban growth. Since film sound technology was one of these inventions, an exploration of urban noise in early sound films seems crucial. The sounds of modern city life were characterized by simultaneousness and a fast-paced rhythm, and sound film, in particular, offered an apt sonic representation of these modern realities.

Today, Mario Camerini's *Gli uomini, che mascalzoni …* (*What Scoundrels Men Are!*) (1932) is mostly known for its hit song "Parlami d'amore, Mariù," sung by the famous actor and later neorealist filmmaker Vittorio De Sica. Various scholars have examined Camerini's importance for Italian filmmaking during the interwar period as one of the most successful commercial filmmakers of the time.[4] However, little attention has been paid to the expressive meaning and political implications of noise in his works. In this chapter, I argue that the use of urban noise in early Italian sound films such as *Gli uomini* was not a mere backdrop but should be seen as a sonic reflection of Italian modernity and modernization under fascism. Rather than promoting fascist ideology, however, the almost documentary portrayal of Milan and its soundscape was part of a realist aesthetic that acknowledged working-class everyday life and concerns, thus anticipating 1940s Italian neorealism.

Milan under Mussolini

Metropolitan areas played a vital role for economic and societal growth during the global interwar period. Cities such as New York, Paris, London, Berlin, or Milan were focused on restructuring their urban architecture. Serving as a cornerstone for economic models and industrial infrastructure, they became the subject of politicized reasoning. Milan was no exception: Benito Mussolini's vision of "Grande Milano" was to restructure the city centre in

2 See, e.g., Green, "From Noise: Blurring the Boundaries of the Soundtrack," 17–32.

3 Eisenstadt, "Multiple Modernities."

4 See, e.g., Ricci, *Cinema & Fascism*, 145.

order to emphasize its economic and cultural importance.[5] Mussolini also initiated the expansion of the motorway and railroad networks, because it offered a key opportunity to advertise the modern economy and power of fascism. During the 1920s and 1930s, these fascist aspirations culminated in several propositions for railroads and motorways to be built all over Europe, connecting to Italy as a point of departure and return. As Massimo Moraglio points out, the "Italian motorway projects were part of a wider plan in which railways, aviation, and bicycles were used to strengthen a vision of modernity within fascist self-representations, giving rise to ideologies of speed and technological nationalism."[6] Railways and motorways also allowed for a more efficient and controlled mode of transporting goods. Due to Mussolini's preference for motorways and truck transportation, the construction of (truck) motorways between Genoa and Milan as well as Turin was eagerly planned during the early 1930s.[7]

While transportation was essential for fascist ideas of modernization, the city itself also played a crucial role in terms of societal control and symbolic emphasis: on the one hand, the construction of new buildings marked the fascist aesthetic of modernity, on the other hand, accenting the old, historic part of Milan emphasized the traditional values of the nation's architectural heritage.[8] But even though Milan's population grew rapidly around 1930, fascist ideology did not necessarily favour this tendency in terms of societal development. As Lucy Maulsby points out, "low birth rates, high mortality, growing unemployment, and continued civil unrest in industrial centres justified the regime's distrust of dense urban areas and resulted in new laws intended to stem the growth of cities." Similar to Nazi Germany, Italy's fascist government used the symbolism of the metropolis for the admonition that city life entailed potential "moral, social, economic, and cultural evils."[9] The juxtaposition of traditional values and modernity, countryside and metropolis, shaped fascist politics and their definition of progress as a controlled form of economic and cultural advancement, with conservative and traditionalist values at the core of a totalitarian system. This contrast between tradition and progressivism was often addressed in films during this period.[10]

5 Maulsby, *Fascism, Architecture and the Claiming of Modern Milan*, 31.

6 Moraglio, *Driving Modernity*, 2.

7 Ibid., 131–35.

8 Maulsby, *Fascism, Architecture and the Claiming of Modern Milan*, 33–34.

9 Ibid., 33.

10 The regime's motto of "ruralizzare l'Italia" was mostly addressed in films from the 1920s until the mid-1930s. For example, Guido Brignone's *Passaporto rosso* (*Red Passport*) (1935) or Mario

The Italian Film Industry during the Time of Fascism

During the 1930s, the Italian film industry was financially supported by Mussolini, who wanted to make use of the visual power and potential of film by promoting national productions and reducing the dominance of foreign films, especially Hollywood films.[11] The cinema had become one of the most beloved forms of entertainment, and the government wanted to further the production of entertaining films meant to distract from reality, as well as of educational films that served as propagandistic tools.[12] The regime's interest in the film industry resulted in new legislation in 1931, which meant that, "for the first time in Europe, a state invested a non-recoverable capital in the entertainment industry."[13] Furthermore, the founding of LUCA Institute (L'Unione Cinematografica Educativa) in 1924 and Cinecittà in 1937 (founded by Benito and his son Vittorio Mussolini as well as Luigi Freddi) enhanced the financial and studio-based potentials of filmmaking.[14] However, all of these state initiatives were built on fascist censorship and entailed the control of artistic output, while maximizing its propagandistic promise and its potential to deflect from negative conditions in the country leading up to the Second World War.[15] That is not to say that Italy's propagandistic intentions were on a par with Nazi Germany's: Luigi Freddi's influence and control over the film industry was not as autocratic as Joseph Goebbel's censorship and propagandistic efforts in the German film industry – in Italy, it was more about overseeing all of the industry and mostly controlling the distribution and exhibition than about rigorous control of the production process.[16]

The coming of sound offered an additional opportunity for censorship by the fascist regime: it allowed for foreign films with 'critical' content to be adapted and censored by way of dubbing.[17] Still, even dubbing was not always a feasible means of censorship for all films. In the case of the American gangster film, for instance, the fascist regime denounced its narrative subtext as a 'lesson learned' on the evil potentials of foreign urban

Camerini's film *Il grande appello* (*The Last Roll-Call*) (1936) expressed this ruralist tendency in their narratives (Brunetta, *The History of Italian Cinema*, 69–70, 96).

11 Brunetta, *The History of Italian Cinema*, 67.

12 Ibid., 74.

13 Ibid., 68.

14 Ibid., 90.

15 Ibid., 70.

16 Reich, "Mussolini at the Movies," 16–17.

17 Ibid., 13.

modernity getting out of hand.[18] Compared to international films, very few Italian films were censored: between 1930 and 1944, only *Ragazzo* (*Boy*) (1934, dir. Ivo Perilli) was prevented from release. Apart from that, Camerini had to remove all joking remarks on dictators and unfair taxes from his *Il capello a tre punte* (*Three Cornered Hat*) (1935), and some other films were taken out of distribution, among them Goffredo Alessandrini's *Noi vivi/Addio Kira!* (*We the Living/Addio Kira!*) (1942) for sympathizing with communism and Luchino Visconti's *Ossessione* (*Obsession*) (1943) for its sexual and political allusions.[19]

During the period, not all film institutions supported fascist ideology. Most notably, the Centro Sperimentale di Cinematografia in Rome, founded by Luigi Chiarini and Umberto Barbaro, became a centre for many antifascist intellectuals.[20] This seemingly unusual co-existence of fascist and liberal or even anti-fascist creatives was not uncommon for Italy, as Gian Piero Brunetta observes. The films resulting from these politically diverse institutions, along with the various production modes, laid bare political contradictions within the film industry of the fascist era. Instead of suppressing one political point of view – the anti-fascist one – the Italian film industry tolerated the creative, productive, and artistic co-existence of fascists and antifascists, allowing them to "live side by side and collaborate in a free ideological and creative trade zone."[21] This was partly due to the fact that the fascist regime did not prioritize fiction film as a propaganda tool – in contrast to the Luce newsreel, which was a controlled propaganda outlet. Fiction films were granted more liberty, in line with Italian minister Giuseppe Bottai's observation that they were not suitable to educate.[22] Comedies in particular – especially the Hollywood-influenced *telefoni bianchi* films – served an escapist function, distracting the people from the economic and political reality.[23]

During the advent of sound, these liberties allowed for experimentation as well as advances in sound recording and sound editing. International exchange was not only encouraged on the fascist side, in order to strengthen the Italian cinema,[24] but also by liberal and/or anti-fascist filmmakers, who

18 Ricci, *Cinema & Fascism*, 155.

19 Reich, "Mussolini at the Movies," 13.

20 Brunetta, *The History of Italian Cinema*, 72.

21 Ibid., 68.

22 Ibid., 69.

23 Zagarrio, "The First Comedy, Italian Style: Blasetti, Camerini, De Sica," 58.

24 For example, Mussolini assigned the direction of *Acciaio* (1933) to Walter Ruttmann (Haaland, *Italian Neorealist Cinema*, 8).

were greatly influenced by the European avant-garde in their approach to sound, montage, and aesthetics.[25] Brunetta describes the development of the 1930s production landscape as a "three-pronged fork in the road." The first of these trends was led by Barbaro and the young cineastes of the Centro Sperimentale di Cinematografia and the journal *Cinema*. Their filmmaking was heavily influenced by Russian film theory and montage. The second one was shaped by filmmakers such as Alessandro Blasetti, who dedicated himself to celebrating "the everyman sound of fascism." And lastly, there was the path Mario Camerini took: exploring urban space as well as its outskirts and reflecting on the geography of "collective petty desires," especially of the working class – themes that are also present in *Gli uomini*.[26]

Mario Camerini and *Gli uomini, che mascalzoni ...*

Before *Gli uomini*, Camerini made *Figaro e la sua gran giornata* (*Figaro and His Great Day*) (1931), one of the first sound films in Italy. At the time, Camerini was hired to work in Paramount's Joinville studios in France, directing a film adaption of Joseph Conrad's novel *Victory*, which became *La riva dei bruti* (*Wild Coast*) (1931), the Italian-language version of *Dangerous Paradise* (1930) (dir. William A. Wellman). As Antonella Sista has shown, Camerini didn't just learn how to write a Hollywood script with a happy ending during his stay; he also became an Italian pioneer in sound editing and post-synchronization, no longer dependent on using only direct sound. In *Gli uomini*, these new techniques enabled Camerini to shoot on location in Milan with "the fast dynamics of a silent movie," adding sound later.[27] Sound editing and post-synchronization indeed allowed for a new 'dialogue,' as it were, between image and sound, which would influence the aesthetics of Italian sound film to come.

Camerini's films were neither openly critical of the fascist regime nor did they entail any propagandistic agenda.[28] His film style was characterized

25 Brunetta, *The History of Italian Cinema*, 79.

26 Ibid., 100.

27 Sisto, *Film Sound in Italy*, 42–43.

28 Only Camerini's comedy *Il cappello a tre punte* (1935) – Carlo Celli calls it "antistatist" – was subject to fascist censorship. It was almost banned by Mussolini and only released in a shorter version due to the "intervention of Minister Alessandro Pavolini and severe cuts." Most of Camerini's other films shifted from the critique of autocratic power to the depiction of the leisure classes, only indirectly alluding to a critique aimed at fascism and Mussolini (Celli, "The Legacy of Mario Camerini" 12–13). Films such as *Kif tebbi* (1928), *Rotaie* (*Rails*) (1931), *Il grande*

by realist tendencies, which subtly revealed the effects of autocratic power as well as working-class concerns during the fascist regime, while adhering to fascist regulations with Hollywood-esque happy endings. This mixture allowed Camerini to appeal to a wide, commercial audience in Italy as well as transnationally.[29] One of the central themes in his movies is the questioning of social identity. Often, the protagonist is mistaken for somebody else or temporarily removed from their social environment.[30] Camerini's films can be considered a prelude to post-war neorealism, which they strongly influenced in terms of narrative structures and aesthetics. It is thus no surprise that Vittorio De Sica – one of the most prominent filmmakers of neorealism – started his film career with Camerini.

Gli uomini's plot revolves around a typical love conflict. Bruno (De Sica), a chauffeur, falls in love with a taxi driver's daughter, Mariuccia (Lia Franca). Only equipped with a bicycle and a low salary, he tries to impress her with a borrowed car and alleged wealth. After some back and forth, the story ends with a happy ending: Bruno makes up with Mariuccia and even gets her father's blessing for their marriage. Typical for Camerini, the film shows everyday life, focusing mostly on blue- and white-collar working-class individuals, who try to find their way in the urban landscape, envisioning their constant self-improvement.[31] While the portrayal of the everyday, especially the sincere look at working-class concerns, did not reflect Mussolini's vision of the Italian individual, the positive affirmation of self-improvement within modern city life did.[32]

If there is a tendency toward cinematic realism in Camerini's films during the fascist era, sound, particularly urban noise, played a major part in this, especially the way it interacted with the images. *Gli uomini* begins with a scene reminiscent of many city symphonies of the 1920s and early 1930s: we see several shots accompanied by nondiegetic music, introducing us to the wake-up routine of a seemingly typical morning in Milan. The first shot is through the window of a department store, with someone outside pulling up the shutters halfway, before the entire window and a vista of the Milan Cathedral is revealed. This first shot thus functions like a theatre curtain that

appello (*The Last Roll-Call*) (1936), or *T'amerò sempre* (*I Will Love You Always*) (1933 and remade by Camerini in 1943) addressed "themes and causes of the regime" like colonialism, labour, and a "natalist campaign" with a certain conservatism regarding family values and acceptance of one's social status (Ben-Ghiat, *Fascist Modernities*, 84; 153).

29 Ricci, *Cinema & Fascism*, 153.

30 Ibid., 146.

31 Konewko, *Neorealism and the "New" Italy*, 16.

32 Brunetta, *The History of Italian Cinema*, 82.

Figure 15.1: Car engines and honking can be heard when taxis come and go in *Gli uomini, che mascalzoni …* (1932, dir. Mario Camerini).

opens to reveal the big city theatre. It is followed by several other on-location shots of Milanese architecture, streets, and a kiosk, setting the urban milieu in which the story is about to unfold. The switch from nondiegetic music to diegetic sound occurs once the story begins: the loud noise of car engines and honking, along with various long shots of taxis leaving or arriving back at their office, introduce the working-class environment (Fig. 15.1). We see taxi driver Tadino, Mariuccia's father and a single parent, as he ends his all-night shift with his usual stop by the café around the corner, before he goes home and wakes up his daughter, who needs to get ready for her work at a perfume shop. The scene back home is again accompanied by nondiegetic music. Interestingly, noises in this opening sequence are only audible in the taxi office and the café – in the former, we hear a collage of loud motor and honking noises, in the latter, the distinct clattering of dishes and glasses, indicating the soundscape of daily urban routines and working-class labour. Car noises, in particular, resonate with an economic concern under fascism: Mussolini's nationalistic policy was to strengthen Italy's automobile industry, specifically the Fiat company, one of the main producers of cars at the time.[33]

33 Moraglio, *Driving Modernity*, 125–26.

Noises in *Gli uomini*, then, document the urban milieu of Milan under fascism. But they also advance the narrative at the film's core: the love story between Mariuccia and Bruno. Furthermore, they underpin class differences, most prominently the one between Bruno and Mariuccia: Bruno only owns a bicycle and aspires to own a car to win over Mariuccia, who, he assumes, is not working class. It is the borrowed automobile that allows Bruno to take Mariuccia on a road trip to the countryside outside of Milan, which contrasts with the vibrant, urban environment we saw earlier. This contrast is mostly emphasized through a romanticized depiction of leisure time and the absence of city noise, especially in the scene of Mariuccia and Bruno dancing and singing "Parlami d'amore, Mariù."[34] A few times, sudden off-screen honking, caused by children playing with Bruno's car, auditively contrasts with the rural setting, as if to bring Bruno back to reality and his status quo, where he is the chauffeur rather than the owner of the car.

It is because of an unlucky coincidence involving his employer that Bruno must suddenly leave Mariuccia behind at the restaurant. The following sequence shows the Milano-Laghi highway, and the noise carries the narrative, alluding to the infrastructure put in place by the regime and the middle-class concerns existing outside of the fictional world.[35] Bruno is seen driving on this highway three times – once leaving town with Mariuccia, then on the way back to Milan with the owner of the car, and finally, when he hastily tries to return to Mariuccia to pick her up. All three scenes are marked by nondiegetic music and a fast-cut, rhythmic montage of close-ups of wheels, the highway, and cars driving by. Reminiscent of Walter Ruttmann's rhythmic editing in *Berlin. Die Sinfonie der Großstadt* (*Berlin: Symphony of A Great City*) (1927), they create a disorientating spatiality, culminating in the last shots of the third scene, with Bruno's car scraping a wagon and crashing into an advertisement sign. It is only then that the music stops, and we suddenly hear the rumbling sounds of a car crashing, a sign falling down, and a man screaming. The sound emphasizes the contrast between the countryside and the rhythms of urban daily life, when Bruno, being an

34 The contrasting of the countryside with urban life was a crucial aspect during the fascist period and of great interest to the government. Even if there is no directly propagandistic aspect to the contrast in *Gli uomini*, the film romanticizes the countryside to a certain degree. There were Italian films with a more propagandistic ideology regarding rural life, such as Blasetti's *Terra madre* (*Mother Earth*) (1931), which almost functioned as an "advertisement for the regime's 'revival' of popular traditions" and "glorified rural lifestyles at a time of low grain output and exodus from the countryside" (Ben-Ghiat, *Fascist Modernities*, 81).

35 This highway was inaugurated in 1924 by Piero Puricelli, who worked for Mussolini (Moraglio, *Driving Modernity*, 54).

urban citizen, almost knocks down a farmer with his wagon on the side of the road. Furthermore, the noise of the car crash alludes to a social reality outside of this film. One of the problems of highways during the 1920s and early 1930s was the absence of a line separating the two lanes. This was due to the roads' infrequent use and lack of traffic, and it increased the potential risk of cars crashing.[36] In addition, the liberty of mostly empty highways led to increased speeding, which is also emphasized by Bruno's hastiness in the film.

The vast highways and implied speeding are contrasted with another scene, showing a traffic jam on the Piazza Guglielmo Oberdan. We see Bruno in his newly found job as a chauffeur working for an elderly man, who happens to have met Mariuccia on the streets and is clearly interested in her. The scene ends with Bruno bursting into anger and leaving the taxi behind in the middle of the traffic jam. Here, the collage-like noises of the traffic announce what we see later in several long shots of the Piazza as well as the Corso Vercelli, when Bruno walks into the middle of the traffic jam, prompting audible honks from cars now stuck behind the abandoned automobile (Fig. 15.2). In this scene, the honking and motor sounds extend the visual density to the aural dimension, creating an urban space that goes beyond the limits of the frame.

Another important location for the representation of noise in this film is the amusement park where Bruno and Mariuccia spend an evening. The crowd noise here is introduced with medium shots of electric lights and signs, serving as an emblem of the transformation of urban space through electricity. When the crowd is finally shown in the shot, its noise has already created a sense of the vastness of the park beyond the screen.[37] Just around the corner, at Mariuccia's new workplace, noises of a perfume box, a machine that produces caramel, and loudspeakers transmitting music – all shown in close-ups – add to the atmosphere of the setting in which the interactions between Bruno and Mariuccia take place.

The Milan Trade Fair (Fiera Campionaria), where both Mariuccia and later Bruno find new work, is another crucial site in this film. As Ruth Ben-Ghiat has stated, large exhibitions, such as the Fiera Campionaria, played a crucial role in Italian fascism and were components "of a comprehensive politics of exhibition(ism)" during the fascist period. The government used these to "communicate particular visions of social organization" and to "substantiate

36 Ibid., 55–57.

37 The amusement park is a common setting in movies of the time, reflecting leisure and social life in the modern city, e.g., in the Hollywood production *Lonesome* (1928, dir. Paul Fejos).

Figure 15.2: Collage-like traffic noises when Bruno (Vittorio de Sica) walks on the streets of Milan.

their power," that is, as an opportunity to exhibit Italy's economic and cultural power in Europe and to fund future Italian inventions through commissions.[38] The exhibitor's hall where Bruno works is shown in several top shots (Fig. 15.3), while various loud machines can be heard and then also seen in a montage of close-ups, conveying the force and power of technological innovation under fascism. At one point, there is a close-up of the noisy water fountain where Bruno is standing to advertise the electric power of a water pump. Its noise almost drowns out Bruno's voice, which is only audible through the megaphone strapped to his head.[39] The loud noise of a siren at the end of the sequence announces the end of labour. Thus, the whole scene comes close to a 'symphony of noises,' reflecting modern labour as well as showcasing new technological, electrical inventions, thereby documenting the noisy modernity of this Milanese exhibition hall.

Besides the socio-politically and spatially charged dimension of filmic noise, Camerini's extensive sound editing also entails more playful uses of sound, for example to accentuate comedic elements, or to create audio-visual

38 Ben-Ghiat, *Fascist Modernities*, 35–36.

39 As Ben-Ghiat points out, whenever Mariuccia and Bruno are together in an urban environment, "the city's voice (tram bells, taxi horns, and crowds) frequently drowns out their own" (Ben-Ghiat, *Fascist Modernities*, 85).

Figure 15.3: Machine noises during top shots of the *Fiera Campionaria* (Milan Trade Fair).

Figure 15.4: Simultaneous visual and audio dissolve in *Gli uomini, che mascalzoni ….*

connections between shots. The comedic aspect is shown in an early scene of Mariuccia taking the tram to work, while Bruno follows her on his bicycle. The only sounds, apart from music, are traffic noises and the tram bells, as well as the noise of water splashing onto Bruno as he drives right through a puddle of water. The audio-visual connections between shots are most evident in one of the perfume shop scenes, where Mariuccia sprays perfume from a bottle into the air: as the accompanying sound fades into a similar sound, the shot dissolves into one of Bruno spraying polish on a car in his garage (Fig. 15.4). Here, Camerini creates an aural connection, or sound bridge, between two objects and urban labour spaces. The sequence is paradigmatic of how noise is no longer a mere adjunct to the visual language of film but an equal means of aesthetic expression. Camerini's use of sound underlines this equality.

Just as the film began with cars, it ends with one. With the happy ending in place, Bruno sits in his cab with Mariuccia's father: we see a close-up of the spinning front wheel (not by accident, it's a Fiat). Instead of car noise, we hear an orchestral score this time, supporting the Hollywood-esque tone of the ending, yet the image of the spinning wheel evokes the car noises heard earlier in the film.[40] Arguably, the wheel and thus the automobile is the film's *actual* protagonist, which quite literally drives the story line. In this sense, the film is reminiscent of other urban comedies, such as Julien Duvivier's *Âllo Berlin? Ici Paris!* (*Here's Berlin*) (1932), where the telephone is the driving force behind the story. In *Gli uomini*, the spinning wheel becomes a visual (and implicitly auditory) metaphor for the struggling working classes and simultaneously for the prevailing fascist ideology, telling spectators that love, family, and consumerism can eventually overcome economic and societal hardship.

The noises in the film ultimately echo this ambivalence. While they document and reflect the technological power and infrastructure of Milan during the fascist period, they also lay bare the individual's struggles in an urban environment. *Gli uomini* thus allowed for modern everyday life in the city (driving a taxi, working in a perfume shop, at a vendor's or in an exhibition hall) and for working-class desires (driving and owning an automobile, being employed and climbing up the social ladder) to be seen as well as heard.

40 While the automobile was part of the fascist ideal of transportation (besides the railway), close-ups of bicycles later became synonymous with neorealist films, most famously through *Ladri di biciclette* (1948, dir. Vittorio De Sica). In this film, the close-ups of Bruno's bicycle contrast with the noise and close-ups of the car wheels, symbolizing class differences.

Bibliography

Ben-Ghiat, Ruth. *Fascist Modernities: Italy, 1922–1945*. Berkeley, Los Angeles: University of California Press, 2001.

Ben-Ghiat, Ruth. "The Italian Cinema and the Italian Working Class." *International Labor and Working-Class History*, no. 59 (Spring 2001): 26–51.

Bertellini, Giorgio. "Dubbing *L'Arte Muta*: Poetic Layerings around Italian Cinema's Transition to Sound." In *Re-Viewing Fascism: Italian Cinema, 1922–1943*, edited by Jacqueline Reich and Piero Garofalo, 30–82. Bloomington: Indiana University Press, 2002.

Brunetta, Gian Piero. *The History of Italian Cinema: A Guide to Italian Film from Its Origins to the Twenty-First Century*. Translated by Jeremy Parzen. Princeton, NJ: Princeton University Press, 2009.

Celli, Carlo. "The Legacy of Mario Camerini in Vittorio De Sica's 'The Bicycle Thief' (1948)." *Cinema Journal* 40, no. 4 (Summer 2001): 3–17.

Dotto, Simone. "How We Learned to Listen to Noise. Theoretical Issues, Practical Problems: Sonic and Media Materiality in Italian Early Sound Film (1929–1935)." *The New Soundtrack* 7, no. 2 (2017): 121–136.

Eisenstadt, S. N. "Multiple Modernities." *Daedalus* 129, no. 1 (Winter 2000): 1–29.

Green, Liz. "From Noise: Blurring the Boundaries of the Soundtrack." In *The Palgrave Handbook of Sound Design and Music in Screen Media*, edited by Liz Greene and Danijela Kulezic-Wilson, 17–32. London: Palgrave Macmillan, 2016.

Haaland, Torunn. *Italian Neorealist Cinema*. Edinburgh: Edinburgh University Press, 2012.

Konewko, Simonetta Milli. *Neorealism and the "New" Italy: Compassion in the Development of Italian Identity*. London: Palgrave Macmillan, 2016.

Maulsby, Lucy M. *Fascism, Architecture and the Claiming of Modern Milan, 1922–1943*. Toronto: University of Toronto Press, 2014.

Moraglio, Massimo. *Driving Modernity: Technology, Experts, Politics, and Fascist Motorways, 1922–1943*. New York: Berghahn, 2017.

Pitassio, Francesco. "Technophobia and Italian Film Theory in the Interwar Period." In *Techné/Technology: Researching Cinema and Media Technologies – Their Development, Use, and Impact*, edited by Annie van den Oever, 185–95. Amsterdam: Amsterdam University Press, 2014.

Reich, Jacqueline. "Mussolini at the Movies: Fascism, Film, and Culture." In *Re-Viewing Fascism: Italian Cinema, 1922–1943*, edited by Jacqueline Reich and Piero Garofalo, 3–29. Bloomington: Indiana University Press, 2002.

Ricci, Steven. *Cinema and Fascism: Italian Film and Society, 1922–1943*. Berkeley: University of California Press, 2008.

Sisto, Antonella C. *Film Sound in Italy: Listening to the Screen*. New York: Palgrave Macmillan, 2014.
Zagarrio, Vito. "The First Comedy, Italian Style: Blasetti, Camerini, De Sica." In *The Italian Cinema Book*, edited by Peter Bondanella, 58–65. London: British Film Institute, 2014.

About the Author

Nadine Soraya Vafi is a PhD candidate at the Department of Film Studies, University of Zurich where she also teaches. Her thesis focuses on the historical contextualization and aesthetics of urban noise in early sound films during the Interwar Period. She was awarded the UZH Doc.Mobility as well as the GRC Travel Grant for her research fellowships at Waseda University and at Università degli Studi di Udine. She worked for international film festivals (e.g., Locarno, Raindance) and was the artistic co-director of the Architektur FilmTage Zürich.

General Bibliography

Abel, Richard, ed. *French Film Theory and Criticism, 1907–1939. Vol. II: 1929–1939*. Princeton, NJ: Princeton University Press, 1988.

Advertisement for *Enough to Eat? The Nutrition Film* (1936). *World Film News* 1, no. 10 (January 1937).

Aitken, Ian. *Alberto Cavalcanti: Realism, Surrealism and National Cinema*. Trowbridge: Flick Books, 2000.

Akhushkov, Shamil, ed. *Kak voznik i stavilsya Vstrechnyi*. Moscow: Kinofotoizdat, 1935.

Alpers, Boris. *Dnevnik kinokritika: 1928–1937*. Moscow: fond Novoe tysiacheletie, 1995.

Altman, Rick. *The American Film Musical*. Bloomington: Indiana University Press, 1987.

Altman, Rick. "Establishing Sound." *CiNéMAS* 24, no. 1 (2013): 19–33.

Altman, Rick. "The Evolution of Sound Technology." In *Film Sound: Theory and Practice,* edited by Elisabeth Weis and John Belton, 44–53. New York: Columbia University Press, 1985.

Altman, Rick. "Penser l'histoire du cinéma autrement: un modèle de crise." *Vingtième Siècle, revue d'histoire* 46 (1995): 65–74.

Altman, Rick. "A Semantic/Syntactic Approach to Film Genre." *Cinema Journal* 23, no. 3 (1984): 6–18.

Altman, Rick. *Silent Film Sound*. New York: Columbia University Press, 2004.

Altman, Rick, McGraw Jones, and Sonia Tatroe. "Inventing the Cinema Soundtrack: Hollywood's Multiplane Sound System." In *Music and Cinema*, edited by James Buhler, Caryl Flinn, and David Neumeyer, 339–59. Hanover: Wesleyan University Press, 2000.

Arnheim, Rudolf. *Film as Art.* Berkeley: University of California Press, 1957.

Arnheim, Rudolf. "A New Laocoön: Artistic Composites and the Talking Film" [1938]. In *Film Sound: Theory and Practice*, edited by Elisabeth Weis and John Belton, 112–15. New York: Columbia University Press, 1985.

Arnheim, Rudolf. "Partly Expensive, Partly Good" [1931]. In *Film Essays and Criticism*. Translated by Brenda Benthien, 173–79. Madison: University of Wisconsin Press, 1997.

Arnheim, Rudolf. *Radio* [1933]. Translated by Margaret Ludwig and Herbert Read. London: Faber & Faber, 1936.

Arnheim, Rudolf. "Die Zukunft des Tonfilms" [1934]. *Montage AV* 9.2 (2000): 18–32.

Arnoux, Alexandre. "Il peut se créer un art neuf." *Pour Vous*, November 29, 1928.

Arnoux, Alexandre. "La bataille des langages et le film parlant." *Pour Vous*, September 5, 1929.

Aros. "Der Turmbau zu Babel: Ein Interview mit Generaldirektor Dent aus London." *Der Kinematograph*, May 21, 1931.

Arrowsmith Production Reports, Samuel Goldwyn Papers, Special Collections, Margaret Herrick Library, Academy of Motion Picture Arts and Sciences, Beverly Hills, California.

Asper, Helmut G. "'The Top Picture to Top All Pictures': Walter Reischs Drehbuch für *F.P.1 antwortet nicht*." In *Walter Reisch. Film schreiben*, edited by Günter Kreen, 120–37. Wien: Filmarchiv Austria, 2004.

Aveline, Claude. "Présentation – Muet sonore ou parlant? La Dernière compagnie – Trader Horn." *Chroniques d'un cinéphile*, 17–20. Paris: Séguier, 1994.

B. [= William Bernard]. "Les films de la semaine." *La Tribune de Genève*, September 16, 1931.

Balázs, Béla. *Early Film Theory: "Visible Man" and "The Spirit of Film"*, edited and translated by Erica Carter. New York: Berghahn, 2010.

Balázs, Béla. "Radio Drama" [1924]. Translated by Russell Stockman. *October* 115 (Winter 2006): 47–48.

Balázs, Béla. "Theory of the Film: Sound." In *Film Sound: Theory and Practice*, edited by Elisabeth Weis and John Belton, 116–25. New York: Columbia University Press, 1985.

Barrier, Michael. *The Animated Man: A Life of Walt Disney*. Berkeley: University of California Press, 2007.

Barnier, Martin. *En route vers le parlant: Histoire d'une évolution technologique, économique et esthétique du cinéma (1926–1934)*. Liège: CEFAL, 2002.

Barnier, Martin. "Versions multiples et langues en Europe." *Mise au Point*, no. 5 (2013). Accessed March 5, 2023, https://journals.openedition.org/map/1490.

Barrier, Michael. *Hollywood Cartoons: American Animation in Its Golden Age*. New York: Oxford University Press, 1999.

Barrier, Michael, Milton Gray, and Bob Clampett. "Interview with Wilfred Jackson (1973)." *MichaelBarrier.com*, July 31, 2015. Accessed February 2, 2023. http://www.michaelbarrier.com/Interviews/Jackson1973/Jackson1973.html.

Barrios, Richard. *A Song in the Dark: The Birth of the Musical Film*. New York: Oxford University Press, 1995.

Bauer, Alfred. *Deutscher Spielfilm-Almanach 1929–1950*. Berlin: Filmblätter Verlag, 1950.

Beck, Jay, and Tony Grajeda, eds. *Lowering the Boom: Critical Studies in Film Sound*. Urbana: University of Illinois Press, 2008.

Ben-Ghiat, Ruth. *Fascist Modernities: Italy, 1922–1945*. Berkeley: University of California Press, 2001.

Ben-Ghiat, Ruth. "The Italian Cinema and the Italian Working Class." *International Labor and Working-Class History*, no. 59 (Spring 2001): 26–51.

Benjamin, Walter. "The Work of Art in the Age of Mechanical Reproduction" [1936]. In *Illuminations*. Translated by Harry Zohn, 217–51. New York: Schocken Books, 1969.

Bennet, Christophe. *La musique à la radio dans les années trente*. Paris: L'Harmattan, 2010.

Bennet, Christophe. "Pianist Kartun and Composer Erwin: From Celebrity to Their Internment." Accessed March 5, 2023, https://halshs.archives-ouvertes.fr/halshs-0114657.

Bertellini, Giorgio. "Dubbing *L'Arte Muta*: Poetic Layerings around Italian Cinema's Transition to Sound." In *Re-Viewing Fascism: Italian Cinema, 1922–1943*, edited by Jacqueline Reich and Piero Garofalo, 30–82. Bloomington: Indiana University Press, 2002.

Billard, Pierre. *L'Âge classique du cinéma français: Du cinéma parlant à la Nouvelle Vague*. Paris: Flammarion, 1995.

Bischoff, F[ritz] Walter. "Hallo! Hier Welle Erdball!" [1928] In *Radio-Kultur in der Weimarer Republik*, edited by Irmela Schneider, 124. Tübingen: Narr, 1984.

Blok, Alexander. *Sobraniye sochinenii*, vol. 7, edited by Vladimir Orlov, Alexander Surkov, Kornei Chukovskii. Leningrad: Gosizdatel'stvo khudozhestvennoi literatury, 1963.

Bogdanovich, Peter. *John Ford*. Rev. ed. Berkeley: University of California Press, 1978. First published 1967.

Bordwell, David. *Narration in the Fiction Film*. Madison: University of Wisconsin Press, 1985.

Bordwell, David, Janet Staiger, and Kristin Thompson. *The Classical Hollywood Cinema: Film Style & Mode of Production to 1960*. New York: Columbia University Press, 1985.

Bordwell, David and Kristin Thompson. *Film Art: An Introduction*. New York: McGraw-Hill, 2019.

Borge, Victor. "Phonetic Punctuation." Accessed February 2, 2023. https://www.youtube.com/watch?v=Qf_TDuhk3No.

Braun, Alfred. *Achtung, Achtung, Hier ist Berlin! Aus der Geschichte des deutschen Rundfunks in Berlin 1923–1932*. Berlin: Haude & Spener, 1968.

Bréan, Samuel, Anne-Lise Weidmann and Jean-François Cornu. "Entretien avec Jean-François Cornu." *Traduire* 230 (2014): 10–21.

Brunetta, Gian Piero. *The History of Italian Cinema: A Guide to Italian Film from Its Origins to the Twenty-First Century*. Translated by Jeremy Parzen. Princeton, NJ: Princeton University Press, 2009.

Buhler, James, David Neumeyer, and Rob Deemer, eds. *Hearing the Movies: Music and Sound in Film History*. New York: Oxford University Press, 2010.

Bulgakowa, Oksana. *Golos kak kul'turnyi fenomen*. Moscow: Novoe Literaturnoe Obozrenie, 2015.

Carnicke, Sharon Marie. "Rethinking 'Stanislavskian' Directing." In *The Great European Stage Directors, Volume 1: Antoine, Stanislavski, Saint Denis*, edited by Peta Tait, 91–111. London: Methuen Drama, 2019.

Cavalcanti, Alberto. "Sound in Films" [1939]. In *Film Sound: Theory and Practice*, edited by Elisabeth Weis and John Belton, 98–111. New York: Columbia University Press, 1985.
Celli, Carlo. "The Legacy of Mario Camerini in Vittorio De Sica's 'The Bicycle Thief' (1948)." *Cinema Journal* 40, no. 4 (Summer 2001): 3–17.
Chaparral. "Dernier cri in Hollywood: Weg mit dem Sprechfilm." *Film-Kurier*, March 12, 1929.
Chion, Michel. *Un art sonore, le cinema: Histoire, esthétique, poétique*. Paris: Cahiers du cinéma, 2003.
Chion, Michel. *Audio-Vision: Sound on the Screen*. Translated by Claudia Gorbman. New York: Columbia University Press, 1994.
Chion, Michel. *Film, a Sound Art*. Translated by Claudia Gorbman. New York: Columbia University Press. 2009.
Chion, Michel. "Glossary: 100 Concepts to Think and Describe Sound Cinema." Accessed February 2, 2023. http://michelchion.com/texts.
Chion, Michel. *The Voice in Cinema*. Translated by Claudia Gorbman. New York: Columbia University Press, 1999.
"Chiqué." *Cinémonde*, February 13, 1930.
Christie, Ian. "Making Sense of Early Soviet Sound." *Inside the Film Factory: New Approaches to Russian and Soviet Cinema*, edited by Richard Taylor and Ian Christie, 176–92. London: Routledge, 1991.
Clair, René. *Cinema Yesterday and Today*, edited by R. C. Dale. New York: Dover Publication, 1972.
Clair, René. "Du théâtre au cinéma." *Anthologie du cinéma*, edited by Marcel Lapierre, 264–70. Paris: La Nouvelle Édition, 1946.
Claus, Horst and Anne Jäckel. "*Der Kongreß tanzt*: Revisited." *CINEMA & Cie*, no. 6 (2005): 76–95.
Conley, Tom. *Film Hieroglyphs: Ruptures in Classical Cinema*. Minneapolis: University of Minnesota Press, 2006.
Cooke, Mervyn. *A History of Film Music*. Cambridge: Cambridge University Press, 2008.
Cooke, Mervyn, and Fiona Ford, eds. *The Cambridge Companion to Film Music*. Cambridge: Cambridge University Press, 2016.
Copeland, Peter. *Sound Recordings*. London: The British Library, 1991.
Corner, John. *The Art of Record*. Manchester: Manchester University Press, 1996.
Cowan, Michael. "Moving Picture Puzzles: Training Urban Perception in the Weimar 'Rebus Films.'" *Screen* 51, no. 3 (Autumn 2010): 197–218.
Cox, Geoffrey. "'There Must Be a Poetry of Sound That None of Us Knows…': Early British Documentary Film and the Prefiguring of Musique Concrète." *Organised Sound* 22, no. 2 (August 2017): 172–86.

Crafton, Donald. *Emile Cohl, Caricature, and Film*. Princeton, NJ: Princeton University Press, 1990.

Crafton, Donald. *Shadow of a Mouse: Performance, Belief, and World-Making in Animation*. Berkeley: University of California Press, 2013.

Crafton, Donald. *The Talkies: American Cinema's Transition to Sound, 1926–1931*. Berkeley: University of California Press, 1997.

Danesi, Marcel. *The Semiotics of Emoji: The Rise of Visual Language in the Age of the Internet*. London: Bloomsbury, 2016.

Davidson, John E., and Theodore F. Rippey, eds. "Early Sound Cinema in the Late Weimar Republic." *Colloquia Germanica* 44, no. 3, (2011): 233–36.

Deleuze, Gilles. *Différence et répétition*. Paris: Presses Universitaires de France, 1968.

Derrida, Jacques. *De la grammatologie*. Paris: Les Éditions de Minuit, 1967.

"Deutschland führt unbestritten in Europa." *Der Kinematograph*, October 6, 1931.

Doré, Claude. "Devant le micro." *Ciné-Miroir*, February 26, 1932.

Dotto, Simone. "How We Learned to Listen to Noise. Theoretical Issues, Practical Problems: Sonic and Media Materiality in Italian Early Sound Film (1929–1935)." *The New Soundtrack* 7, no. 2 (2017): 121–136.

Dr. E. "Filmchronik." *Neue Zürcher Nachrichten*, February 8, 1930.

Durovicova, Natasa. "Introduction." *CINEMA & Cie*, no. 4 (2004): 7–16.

Eades, Mark. "First Draft of Mickey Mouse 'Steamboat Willie' Script on the Auction Block." *Orange County Register*, October 7, 2016. Accessed February 2, 2023. https://www.ocregister.com/2016/10/07/first-draft-of-mickey-mouse-steamboat-willie-script-on-the-auction-block/.

Eidgenössisches Statistisches Amt = Bureau Fédéral de Statistique. *Statistisches Jahrbuch der Schweiz 43. Jahrgang = Annuaire Statistique de la Suisse* 43me *année*. Basel: E. Birkhäuser & Cie, 1935.

Eisenstadt, S. N. "Multiple Modernities." *Daedalus* 129, no. 1 (Winter 2000): 1–29.

Eisenstein, Sergei. "The Cinematographic Principle and the Ideogram." In *Film Form: Essays in Film Theory*, edited and translated by Jay Leyda, 28–44. New York: Harcourt, Brace & World, 1949.

Eisenstein, Sergei. "Dickens, Griffith and the Film Today." In *Film Form: Essays in Film Theory*, edited and translated by Jay Leyda, 195–255. New York: Harvest Book, 1949.

Eisenstein, Sergei. *Eisenstein on Disney*, edited by Jay Leyda. Translated by Alan Upchurch. New York: Methuen, 1988.

Eisenstein, Sergei. *Metod/Die Methode*, edited by Oksana Bulgakowa. Berlin: PotemkinPress, 2009.

Eisenstein, Sergei, Vesvolod Pudovkin, and Grigori Alexandov. "Statement on Sound." Sergei Eisenstein. *Selected Works vol. 1: Writings 1922–34*, edited and translated by Richard Taylor, 113–14, London: I. B. Tauris, 2010.

Eisner, Will. *Comics & Sequential Art.* Exp. ed. Tamarac, FL: Poorhouse Press. 2001.
Eksteins, Modris. "All Quiet on the Western Front." *History Today* 45, no. 11 (1995): 29–34.
Elsaesser, Thomas. "Going 'Live': Body and Voice in Some Early German Sound Film." In *Le son en perspective: Nouvelles recherches / New Perspectives in Sound Studies*, edited by Dominique Nasta and Didier Huvelle, 155–68. Brussels: P.I.E.-Peter Lang, 2004.
Elsaesser, Thomas. "ImpersoNations: National Cinema, Historical Imaginaries and New Cinema Europe." *Mise au Point*, no. 5 (2013). Accessed March 5, 2023, https://journals.openedition.org/map/1480.
Emojipedia "Musical Note." Accessed May 30, 2023. https://emojipedia.org/musical-note/.
Encke, Julia. *Augenblicke der Gefahr. Der Krieg und die Sinne, 1914–1934*. München: Fink, 2006.
F. Ch. [= Freddy Chevalley]. "Devant le film." *La Suisse*, April 7, 1930.
Feyersinger, Erwin. "Diegetic Short Circuits: Metalepsis in Animation," *Animation: An Interdisciplinary Journal* 5, no. 3 (2010–2011): 279–94.
Fielding, Raymond. *The American Newsreel 1911–1967.* Norman: University of Oklahoma Press, 1972.
Fielding, Raymond. *The March of Time, 1935–1951.* Oxford: Oxford University Press, 1977.
Fischbach, Lisa. "Punctuation (1): The Linguistic Side." *Cast*, May 9, 2017. Accessed February 2, 2023. https://articles.c-a-s-t.com/punctuation-1-the-linguistic-side-9ba8eb11fbf2.
Fischer, Lucy. "*Applause*: The Visual and Acoustic Landscape." In *Film Sound: Theory and Practice*, edited by Elisabeth Weis, and John Belton, 232–46. New York: Columbia University Press, 1985.
Fisher, Jaimey. "The Haptic Horrors of War: Towards a Phenomenology of Affect and Emotion in the War Genre in Germany, 1910s to 1950s." *Seminar* 50, no. 1 (2014): 51–68.
Fisher, Jaimey. "Landscapes of Death: Space and the Mobilization Genre in G.W. Pabst's *Westfront 1918* (1930)." In *The Many Faces of Weimar Cinema: Rediscovering Germany's Filmic Legacy*, edited by Christian Rogowski, 268–85. Rochester: Camden House, 2010.
Fleeger, Jennifer. *Sounding American: Hollywood, Opera, and Jazz.* New York: Oxford University Press, 2014.
Forsyth Hardy. "First Principles of Documentary," [1933–1934]. In *Grierson on Documentary*, edited by Forsyth Hardy, 145–56. London: Faber and Faber, 1979.
Forsyth Hardy. "The GPO Gets Sound." *Cinema Quarterly* 2, no. 4 (1934): 215–21.
Forsyth Hardy. "Two Paths to Poetry." *Cinema Quarterly* 3, no. 4 (1935): 194–96.

Freund, Karl. "Just What is Montage." *American Cinematographer* 15, no. 9 (1934): 204; 210.

Friedell, Egon. *A Cultural History of the Modern Age: The Crisis of the European Soul from the Black Death to the World War* [1927–1931]. Translated by Charles F. Atkinson. New York: Knopf, 1930–1932.

Gad, Urban. *Der Film: seine Mittel – seine Ziele* [1919]. Translated by Julia Koppel. Berlin: Schuster & Loeffler, 1921.

Galili, Doron. "Intermedial Thought in Classical Film Theory: Balázs, Arnheim, and Benjamin on Film and Radio." *The Germanic Review* 88, no. 4 (2013): 391–99.

Gardin, Vladimir. "Kak ia rabotal nad obrazom Babchenko." *Sovetskoe kino*, no. 1–2 (1934): 96–100.

Garity, William. "The Production of Animated Cartoons," *Journal of the Society of Motion Picture Engineers* 29, no. 4 (1933): 309–22.

Garncarz, Joseph. "Making Films Comprehensible and Popular Abroad: The Innovative Strategy of Multiple-language Versions." *CINEMA & Cie*, no. 4 (2004): 72–79.

Garncarz, Joseph. *Wechselnde Vorlieben. Über die Filmpräferenzen der Europäer 1896–1939*. Frankfurt am Main: Stroemfeld, 2015.

Gehri, Alfred. "Où va le cinéma?" *Feuille d'avis de Lausanne*, September 19, 1930.

Geisler, Michael. "The Battleground of Modernity: *Westfront 1918* (1930)." In *The Films of G.W. Pabst: An Extraterritorial Cinema*, edited by Eric Rentschler, 91–102. New Brunswick, NJ: Rutgers University Press, 1990.

Genova, "La carrière de John Gilbert est-elle en danger?" *Ciné-Miroir*, March 7, 1930.

Girard, René. "Sous les toits de Paris." *Cinémonde*, May 8, 1930.

Goldmark, Daniel. "Before Willie: Reconsidering Music and the Animated Cartoon of the 1920s." In *Beyond the Soundtrack: Representing Music in Cinema*, edited by Daniel Goldmark, Lawrence Kramer, and Richard D. Leppert, 225–45. Berkeley: University of California Press, 2007.

Goldmark, Daniel. *Tunes For 'Toons: Music and the Hollywood Cartoon*. Berkeley: University of California Press, 2005.

Gosho, Heinosuke. "Madamu to nyōbō." In *Nihon eiga shinario koten zenshu vol. 2*, 48, Tokyo: Kinema Junpōsha, 1966.

Gosho, Heinosuke. "Nihon tōkī no danpenteki jōhō." *Ongaku sekai* 6, no. 6, (1934): 57–62.

Green, Liz. "From Noise: Blurring the Boundaries of the Soundtrack." In *The Palgrave Handbook of Sound Design and Music in Screen Media*, edited by Liz Greene and Danijela Kulezic-Wilson, 17–32. London: Palgrave Macmillan, 2016.

Grierson, John. "Battle for Authenticity" [1939]. In *Grierson on Documentary*, edited by Forsyth Hardy, 215–17. London: Faber & Faber, 1979.

Gunning, Tom. *The Films of Fritz Lang: Allegories of Vision and Modernity*. London: BFI, 2000.

Gunning, Tom. "'From the Bottom of the Sea': Early Film at the Oberhausen Festival." In *Early Cinema Today: The Art of Programming and Live Performance*, edited by Martin Loiperdinger, 37–41. New Barnet: John Libbey, 2011.

Gunning, Tom. "Vachel Lindsay: Theory of Movie Hieroglyphics." In *Thinking in the Dark: Cinema, Theory, Practice*, edited by Murray Pomerance and R. Barton Palmer, 19–30. New Brunswick, NJ: Rutgers University Press, 2016.

Haaland, Torunn. *Italian Neorealist Cinema*. Edinburgh: Edinburgh University Press, 2012.

Hansen, Miriam. "Mass Culture as Hieroglyphic Writing: Adorno, Derrida, Kracauer." *New German Critique*, no. 56 (Spring/Summer 1992): 42–73.

Hanson, Helen. *Hollywood Soundscapes: Film Sound Style, Craft and Production in the Classical Era*. London: Bloomsbury, 2017.

Hardt, Ursula. *From Caligari to California: Erich Pommer's Life in the International Film Wars*. New York: Berghahn, 1996.

HD. "The Mask and the Movietone" [1927]. In *Red Velvet Seat: Women's Writings on the First Fifty Years of Cinema*, edited by Antonia Lant, 205–12. New York: Verso, 2006.

Hd. [= Jean Hennard]. "Les débuts du film sonore en Suisse." *Schweizer Cinema Suisse*, September 1, 1929.

Helmers, Maike. "The Transition from the Silent into the Sound Era: The Innovative Use of Sound in Pabst's Westfront 1918." *Music, Sound, and the Moving Image* 12, no. 2 (2018): 121–39.

Helmholtz, Hermann von. *Die Lehre von den Tonempfindungen als Physiologische Grundlage für die Theorie der Musik*. Hildesheim: G. Olms, 1968.

Henri-Robert, Jacques. "Le film sonore. Ce que le public en pense." *Cinéa*, April 15, 1929.

Horak, Jan-Christopher. "Exilfilm 1933–1945." In *Geschichte des deutschen Films*, edited by Wolfgang Jacobsen, Anton Kaes, and Hans Helmut Prinzler, 101–18. Stuttgart: Metzler, 1993.

Horowitz, Joseph. *"On My Way": The Untold Story of Rouben Mamoulian, George Gershwin, and Porgy and Bess*. New York: W. W. Norton & Company, 2013.

Huhtamo, Erkki and Jussi Parikka, eds. *Media Archaeology: Approaches, Applications, and Implications*. Berkely: University of California Press, 2011.

Ihering, Herbert. "Der Mann, der den Mord beging." *Von Reinhardt bis Brecht: Vier Jahrzehnte Theater und Film, vol. III: 1930–1932*. Berlin: Aufbau-Verlag 1961.

Iida, Shinbi. "Nikkatsu jidaigeki tōkī Tange Sazen." *Kinema junpō*, November 12, 1933.

Ijima, Tadashi. "Furusato." *Kinema junpō*, April 21, 1930.

"Il film internazionale finirà?" *La rivista cinematografica*, January 1, 1930.

Inagaki, Hiroshi. *Nihon eiga no wakaki hibi*. Tokyo: Mainichi shinbunsha, 1978.

Ishigami, Iku. "Funman nihon take." *Kinema junpō*, August 11, 1932.

Itō, Daisuke, Īda Shinbi, Hazumi Tsuneo, Kobayashi Isamu, Kitagawa Fuyuhiko, Kisaragi Bin, Kishi Matsuo, Sugiyama Shizuo. "Itō Daisuke o kakonde: Jidaigeki tōkī zadankai." *Kinema junpō*, January 1, 1934.

Iwamoto, Kenji. "Sound in the Early Japanese Talkies." In *Reframing Japanese Cinema: Authorship, Genre, History*, edited by Arthur Nolletti, Jr. and David Desser, 312–27. Bloomington: Indiana University Press, 1992.

Iwerks, Leslie, and John Kenworthy. *The Hand Behind the Mouse*. New York: Disney Editions, 2001.

J. Ct. [= Jeanne Clouzot]. "De film en film." *Journal de Genève*, April 6, 1930.

J. Nr. [= Jean Nicollier]. "Parmi les films." *Gazette de Lausanne*, April 7, 1930.

J. P. [= Jean Peitrequin]. "Les cinémas." *La Revue*, March 19, 1930.

Jacobs, Lea. *Film Rhythm After Sound: Technology, Music, and Performance*. Berkeley: University of California Press, 2015.

Jacobs, Lea. "The Innovation of Re-Recording in the Hollywood Studios." *Film History* 24, no. 1 (2012): 5–34.

Jäger, Ernst. "Film Kritik: Der Mörder Dimitri Karamasoff." *Film-Kurier*, February 7, 1931.

Jäger, Ernst. "Film-Kritik: Phantome des Glücks." *Film-Kurier*, January 3, 1930.

Janecek, Gerald. *Zaum: The Transrational Poetry of Russian Futurism*. San Diego: San Diego State University Press, 1996.

Jolowicz, Ernst. *Der Rundfunk. Eine psychologische Untersuchung*. Berlin: Max Hesse, 1932.

Kaes, Anton. "29 November 1923: Karl Grune's *Die Straße* Inaugurates 'Street Film,' Foreshadows Film Noir." In *A New History of German Cinema*, edited by Jennifer M. Kapczynski and Michael D. Richardson, 124–28. Rochester, NY: Camden House, 2012.

Kaes, Anton, Nicholas Baer, and Michael Cowan, eds. *The Promise of Cinema: German Film Theory 1907–1933*. Berkeley: University of California Press, 2016.

Kaganovsky, Lilya. *The Voice of Technology: Soviet Cinema's Transition to Sound, 1928–1935*. Bloomington: Indiana University Press, 2018.

Kalganov, Yuri. "Pervye gody Leningradskogo radio." In *Pervye gody sovetskogo muzykal'nogo stroitel'stva: Stat'i: Vospominaniia: Materialy*, 205–30. Leningrad: Sovetskii kompozitor, 1959.

Kane, Brian. *Sound Unseen: Acousmatic Sound in Theory and Practice*. New York: Oxford University Press, 2014.

Kasack, Hermann. "Mikroreportage." In *Radio-Kultur in der Weimarer Republik*, edited by Irmela Schneider, 128–33. Tübingen: Narr, 1984.

Kelly, Andrew. "In the Trenches, on the Screen: World War I on Film." In *The Wiley-Blackwell History of American Film Vol. 1: Origins to 1928*, edited by Cynthia Lucia and Roy Grundmann, Art Simon, 1–16. Hoboken: Wiley-Blackwell, 2011.

Kelly, Michael. "Mimesis." *The Encyclopedia of Aesthetics*. Vol. 3. New York: Oxford University Press, 1998.

Kido, Shiro. *Nihon eigaden: Eiga seisakusha no kiroku*. Tokyo: Bungei shunjū shinsha, 1956.

Kinoshita, Chika. "The Benshi Track: Mizoguchi Kenji's the Downfall of Osen and the Sound Transition." *Cinema Journal* 50, no. 3 (2011): 1–25.

Kinoshita, Chika. *Mise-En-Scène of Desire: The Films of Mizoguchi Kenji*. PhD diss., University of Chicago, 2007.

Kitagawa, Fuyuhiko. "Hanayome no negoto." *Kinema junpō*, February 1, 1933.

Kitagawa, Fuyuhiko. "Jidaieiga ni okeru iwayuru 'wajutsu' nazo." *Kinema junpō*, August 1, 1933.

Kitagawa, Fuyuhiko. "Toritomenonai kanso." *Kinema junpō*, January 12, 1934.

Kiyose, Eijirō. "Tokii jidai-geki no daiyarogu." In *Eiga kagaku kenkyu*, no. 9 (September 1931): 100–101.

Klaus, Ulrich J. *Deutsche Tonfilme*. Berlin: Klaus-Archiv, 1989.

Klimes, Ivan and Pavel Zeman. "Aktualita 1937–1938." *CINEMA & Cie*, no. 6 (2005): 62–70.

Klitzsch, Ludwig. "Der Film als nationaler Kulturfaktor." *Der Kinematograph*, July 20, 1932.

Knebel, Maria. *Vsia zhizn'*. Moscow: VTO, 1967.

Koerber, Martin. "Oskar Messter, Film Pioneer: Early Cinema between Science, Spectacle, and Commerce." In *A Second Life: German Cinema's First Decades*, edited by Thomas Elsaesser and Michael Wedel, 51–61. Amsterdam: Amsterdam University Press, 1996.

Konewko, Simonetta Milli. *Neorealism and the "New" Italy: Compassion in the Development of Italian Identity*. London: Palgrave Macmillan, 2016.

Koszarski, Richard. "The Greatest Film Paramount Ever Made." *Film History* 15, vol. 4 (2003): 436–43.

Kozintsev, Grigori. *Sobranie sochinenii*, edited by Iakov Butovskii and Valentina Kozintseva. Leningrad: Iskusstvo, 1982.

"Kritische Filmschau: Anna Christie." *Deutsche Filmzeitung*, January 23, 1931.

"Kritische Filmschau: Der Schuss im Tonfilmatelier." *Deutsche Filmzeitung*, September 5, 1930.

"Kritische Filmschau: Die Königsloge." *Deutsche Filmzeitung*, July 11, 1930.

"Kritische Filmschau: Die Nacht gehört uns." *Deutsche Filmzeitung*, January 10, 1930.

"Kritische Filmschau: Flieger." *Deutsche Filmzeitung*, May 30, 1930.

"Kritische Filmschau: Liebe im Ring." *Deutsche Filmzeitung*, April 4, 1930.

"Kritische Filmschau: Sag es mit Liedern." *Deutsche Filmzeitung*, July 25, 1930.

"Kritische Filmschau: Sonntag des Lebens." *Deutsche Filmzeitung*, February 27, 1931.

"Kritische Filmschau: Zwei Menschen." *Deutsche Filmzeitung*, January 16, 1931.

L'Herbier, Marcel. "Opinioni di artisti francesi sui films parlanti." *La Rivista cinematografica*, March 30, 1929.

"La Petite Lise." *Cinémonde*, December 11, 1930.

Lastra, James. *Sound Technology and the American Cinema: Perception, Representation, Modernity.* New York: Columbia University Press, 2000.

Leigh, Walter. "The Musician and the Film." *Cinema Quarterly* 3, no. 2 (1935): 70–74.

Leonhard, Rudolf. "Lautsprecher" [1931]. In *Radio-Kultur in der Weimarer Republik*, edited by Irmela Schneider, 56. Tübingen: Narr, 1984.

Leonhard, Rudolf. "Technik und Kunstform" [1924]. In *Radio-Kultur in der Weimarer Republik*, edited by Irmela Schneider 67–72. Tübingen: Narr, 1984.

Lewinsky, Mariann. "The Best Years of Film History: A Hundred Years Ago." In *Early Cinema Today: The Art of Programming and Live Performance*, edited by Martin Loiperdinger, 25–35. New Barnet: John Libbey, 2011.

Lewis, Hannah. *French Musical Culture and the Coming of Sound Cinema.* New York: Oxford University Press, 2018.

Lightman, Herb A. "The Magic of Montage." *American Cinematographer* 30, no. 10 (1949): 381–82.

Lindsay, Vachel. *The Art of the Moving Picture* [1915]. New York: Liveright Publishing, 1970.

London, Kurt. *Film Music: A Summary of the Characteristic Features of Its History, Aesthetics, Technique; and Possible Developments.* London: Faber & Faber, 1936.

London, Kurt. "Wochenschau mit oder ohne Begleitmusik?" *Der Film*, April 15, 1929.

Maletzke, Gerhard. *Psychologie der Massenkommunikation. Theorie und Systematik.* Hamburg: Hans Bredow Institut, 1963.

Mamoulian, Rouben. *The American Film Institute Seminar.* May 27, 1981, RMP.

Mamoulian, Rouben. "The Art of Films." Lecture given to "History of the Motion Picture" class at Columbia University, Museum of Modern Art, December 6, 1939, RMP.

Mamoulian, Rouben. "Common Sense and Camera Angles." *The American Cinematographer* 13, vol. 10 (1932): 8–9, 26.

Mamoulian, Rouben. "The Essence of Theatrical Art." Lecture delivered in Rochester (NY) 1923–1924, RMP.

Mamoulian, Rouben. "An Interview with Rouben Mamoulian." Interview by John A. Gallagher, and Marino A. Amoruco. *The Velvet Light Trap*, no. 19 (1982): 16–22.

Mamoulian, Rouben. "Interview with Rouben Mamoulian." Unpublished interview by William Becvar. March 20–22, 1973, RMP.

Mamoulian, Rouben. Oral History Transcript, Columbia University, New York, December 1958, RMP.

Mamoulian, Rouben. "The Psychology of Sound: The Second in a Series of Papers Challenging the Realistic Conception of Art." July 1938, RMP.

Mamoulian, Rouben. "Rouben Mamoulian: An Exclusive Interview." Interview by William Hare. In *American Classic Screen Interviews*, edited by John C. Tibbetts and James M. Welsh. Lanham, MD: Scarecrow Press, 2010.

Mamoulian, Rouben. "The Use and Abuse of Perambulation." Lecture given to the American Society of Cinematographers, July 19, 1932, RMP.

Mamoulian, Rouben. "Visual Anti-Realism: The First in a Series of Papers Challenging the Realistic Conception of Art." July 1938, RMP.

Mansell, James G. "Neurasthenia, Civilization, and the Sounds of Modern Life: Narratives of Nervous Illness in the Interwar Campaign against Noise." In *Sounds of Modern History: Auditory Cultures in 19th- and 20th-Century Europe*, edited by Daniel Morat, 278–302. New York: Berghahn, 2014.

Marie, Michel. "*Sous les toits de Paris*, film de René Clair." Accessed January 3, 2023. https://www.universalis.fr/encyclopedie/sous-les-toits-de-paris/.

Maulsby, Lucy M. *Fascism, Architecture and the Claiming of Modern Milan, 1922–1943*. Toronto: University of Toronto Press, 2014.

Mayr, Brigitte. "Budapester Charme und Wiener Ambiente. Der Musikfilm im Exil in Österreich und Ungarn 1933–1938." In *Als die Filme singen lernten. Innovation und Tradition im Musikfilm 1928–1938*, edited by Malte Hagener and Jan Hans, 166–78. München: edition text+kritik, 1999.

McCloud, Scott. *Understanding Comics: The Invisible Art*. New York: HarperCollins, 1994.

McGuire, John Thomas. "Filtering and Interpreting The Great War: *All Quiet on the Western Front*, *Journey's End*, *Westfront 1918*, and Their Perspectives on World War I." *Quarterly Review of Film and Video* 33, no. 7 (2016): 667–80.

Merritt, Russell, and J. B. Kaufman. *Walt Disney's Silly Symphonies: A Companion to the Classic Cartoon Series*. Revised and Updated. Glendale, CA: Disney Editions, 2016.

Metz, Christian. *Language and Cinema*. Translated by Donna Jean Umiker-Sebeok. The Hague: Mouton, 1974.

Millard, André. *America on Record: A History of Recorded Sound*. Cambridge: Cambridge University Press, 2005.

Milne, Tom. *Mamoulian*. London: Thames & Hudson, 1969.

Mirbel, Jean de. "Prix de beauté sur la plage de Joinville." *Cinémagazine*, September 20, 1929.

Misono, Kyōhei. *Eiga Torai nanajunen: Eiga Chūshingura*. Tokyo: n.p., 1966.

Mitchell, Greg. *The Campaign of the Century: Upton Sinclair's Race for Governor of California and the Birth of Media Politics*. New York: Random House, 1992.

Mitry, Jean. "Entretien." *Cinématographe*, no. 47 (1979): 18–23.

Miyao, Daisuke. *The Aesthetics of Shadow: Lighting and Japanese Cinema*. Durham: Duke University Press, 2013.

Montagu, Ivor. *With Eisenstein in Hollywood.* New York: International Publishers, 1969.

Moraglio, Massimo. *Driving Modernity: Technology, Experts, Politics, and Fascist Motorways, 1922–1943*. New York: Berghahn, 2017.

Morat, Daniel. "Cheers, Songs, and Marching Sounds: Acoustic Mobilization and Collective Affects at the Beginning of World War I." In *Sounds of Modern History: Auditory Cultures in 19th- and 20th-Century Europe*, edited by Daniel Morat, 177–200. New York: Berghahn, 2014.

Mori, Iwao. *Watashi no geikai henreki*. Tokyo: Seigabo, 1975.

Mori, Iwao, Tanaka Eizō, Hatamoto Shūichi, Kisaragi Bin, Kitamura Komatsu, Mizoguchi Kenji, Mineo Yoshio, Tsutami Jōbu, and Ōno Motomu. "Furusato zadankai." *Eiga ōrai* 6, no. 63 (May 1930): 41–56.

Moure, José. "Du silence au cinéma." In *Voix et média*, edited by Marie Thonon, 24–38. Paris: L'Harmattan, 1999.

Mühl-Benninghaus, Wolfgang. *Das Ringen um den Tonfilm. Strategien der Elektro- und der Filmindustrie in den 20er und 30er Jahren.* Düsseldorf: Droste, 1999.

Müller, Corinna. "Akustik des Krieges. Der Erste Weltkrieg als akustisches Ereignis im frühen Tonfilm." In *Der erste Weltkrieg im Film*, edited by Rainer Rother, Karin Herbst-Messlinger, 90–111. München: edition text + kritik, 2009.

Müller, Corinna. *Vom Stummfilm zum Tonfilm*. München: Wilhelm Fink, 2003.

Nagato Yohei. "The Dawn of the Talkies in Japan: Mizoguchi Kenji's Hometown." In *The Culture of the Sound Image in Prewar Japan*, edited by Michael Raine and Johan Nordström, 183–99. Amsterdam: Amsterdam University Press, 2020.

Nakata, Kameinosuke. "Hitotsu Toki geijutsuron." *Kinema junpō*, January 1, 1934.

Nasta, Dominique, and Didier Huvelle, eds. *Le son en perspective: Nouvelles recherches/New Perspectives in Sound Studies*. Brussels: P.I.E.-Peter Lang, 2004.

Natanson. "Les joies et les ennuis du dialogueur." *Pour Vous*, October 13, 1932.

Newsom, Jon. "'A Sound Idea': Music for Animated Films," *The Quarterly Journal of the Library of Congress* 37, no. 3/4 (Summer/Fall 1980): 279–309.

Nikkatsu kaubushikikaisha. "Kanda Nikkatsu puroguramu." Accessed June 12, 2023, https://www.nikkatsu.com/movie/12752.html.

Nikkatsu kabushikikaisha. *Nikkatsu yonjunenshi*. Tokyo: Nikkatsu, 1952.

Nordström, Johan. "Between Silence and Sound: The Liminal Space of the Japanese 'Sound Version.'" In *Routledge Handbook of Japanese Cinema*, edited by Joanne Bernardi and Shota T. Ogawa, 213–30. London/New York: Routledge, 2021.

Nordström, Johan. "The Image of the Modern Talkie Film Studio: Aesthetics and Technology at P.C.L." In *The Culture of the Sound Image in Prewar Japan*, edited by Michael Raine and Johan Nordström, 157–81, Amsterdam: Amsterdam University Press, 2020.

O'Brien, Charles. *Cinema's Conversion to Sound: Technology and Film Style in France and the U.S.* Bloomington: Indiana University Press, 2005.

O'Brien, Charles. *Movies, Songs, and Electric Sound: Transatlantic Trends*. Bloomington: Indiana University Press, 2019.

O'Rawe, Des. "The Great Secret: Silence, Cinema and Modernism." *Screen* 47, no. 4 (2006): 395–405.

Ōkoku, Fujimoto. "Fugu no Nihon tōkī: Hachisuka Koroku o miru." *Eiga jidai* 6, no. 10 (October, 1929): 20–21.

Olenina, Ana Hedberg. *Psychomotor Aesthetics: Movement and Affect in Modern Literature and Film*. New York: Oxford University Press, 2020.

Olsson, Jan. "Framing Silent Calls: Coming to Cinematographic Terms with Telephony." In *Allegories of Communication: Intermedial Concerns from Cinema to the Digital*, edited by John Fullerton and Jan Olsson, 157–92. Bloomington: Indiana University Press, 2005.

Ōmura, Einosuke. "Ōmura Einosukeshi kikisho (2)." *Kiroku eiga*, no. 279 (February 1986), 7–9.

Ōtsuka, Kyoichi. *Nihon eiga kantokuron*. Tokyo: Eiga hyōronsha, 1937.

P. [= Maurice Porta]. "Dans nos cinémas." *Feuille d'avis de Lausanne*, April 28, 1930.

P. V. [= Pierre Vidoudez]. "Devant les films." *Tribune de Lausanne*, April 20, 1930.

Palitzsch, Otto Alfred. "Der brave Sünder." *Vossische Zeitung*, October 23, 1931.

Piotrovskii, Adrian. *Teatr: Kino: Zhizn'*. Leningrad: Iskusstvo, 1969.

Pitassio, Francesco. "Technophobia and Italian Film Theory in the Interwar Period." In *Techné/Technology: Researching Cinema and Media Technologies – Their Development, Use, and Impact*, edited by Annie van den Oever, 185–95. Amsterdam: Amsterdam University Press, 2014.

Pommer, Erich. "Pommer über den mehrsprachigen Tonfilm." *Der Kinematograph*, September 17, 1930.

Porter, Laraine. "Okay for Sound? The Reception of the Talkies in Britain 1928–1932." *Journal of British Cinema and Television* 17, no. 2 (2019): 212–32.

Posener, Valérie. "To Overtake and Outstrip Hollywood: Early Talking Pictures in the Soviet Union." In *Sound, Speech, Music in Soviet and Post-Soviet Cinema*, edited by Masha Salazkina and Lilya Kaganovsky, 60–80. Bloomington: Indiana University Press, 2013.

Pound, Ezra. *ABC of Reading*. New York: New Directions, 1934.

Pudovkin, Vsevolod. *Sobraniye sochinenii*, edited by Tatiana Zapasnik and Ada Petrovich. Moscow: Iskusstvo, 1974.

R., O. "L'arte muta è minacciata dai progressi del fonocinematografo?" *La rivista cinematografica*, February 28, 1929.

Radek, Karl. "Dve fil'my." *Izvestiia*, April 23, 1931, 4.

Raine, Michael. "No Interpreter, Full Volume: The *Benshi* and the Sound Transition in 1930s Japan." In *The Culture of the Sound Image in Prewar Japan*, edited by

Michael Raine and Johan Nordström, 127–56. Amsterdam: Amsterdam University Press, 2020.

Raine, Michael, and Johan Nordström. *The Culture of the Sound Image in Prewar Japan*. Amsterdam: Amsterdam University Press. 2020.

Régent, Roger and Édouard Sattler. "France, Allemagne, Italie …" *Pour Vous*, September 25, 1930, 6.

Régent, Roger and Édouard Sattler. "Vernon via Séville, Fédor par le Mexique." *Pour Vous*, February 19, 1931.

Reich, Jacqueline. "Mussolini at the Movies: Fascism, Film, and Culture." In *Re-Viewing Fascism: Italian Cinema, 1922–1943*, edited by Jacqueline Reich and Piero Garofalo, 3–29. Bloomington: Indiana University Press, 2002.

Renoir, Jean. *My Life and My Films*. Translated by Norman Denny. London: Collins, 1974.

RGALI (Russian State Archive of Literature and the Arts), Moscow.

Ricci, Steven. *Cinema and Fascism: Italian Film and Society, 1922–1943*. Berkeley: University of California Press, 2008.

Rossholm, Anna Sofia. *Reproducing Languages, Translating Bodies*. Stockholm: Stockholm University, 2006.

Rotha, Paul. *Documentary Diary: An Informal History of the British Documentary Film*. New York: Hill and Wang, 1973.

Rotha, Paul. *Documentary Film*. New York: W. W. Norton & Company, Inc., 1939.

Rotha, Paul. "Films of Fact and Fiction" [1938]. In *A Paul Rotha Reader*, edited by Duncan Petrie and Robert Kruger, 161–68. Exeter: University of Exeter, 1999.

rz. "Berner Filmschau." *Der Bund*, May 2, 1930.

Sakae. "Shineigahyo 'Taii no musume.'" *Asahi Shimbun*, October 31, 1929.

Salt, Barry. "Barry Salt's Database." Accessed February 2, 2022. http://www.cinemetrics.lv/satltdb.php.

Salt, Barry. *Film Style & Technology: History and Analysis*. 3rd ed. London: Starword, 2009. First published 1983.

Sasō, Tsutomu., Nishida, Nobuyoshi, "5. Toki no ujigaki." In *Mizoguchi Kenji: Jōen no hate no onna tachi yo genmu e no riarizumu*, edited by Sasō, Tsutomu and Nishida, Nobuyoshi, 93, Tokyo: Firumuatosha, 1997.

Satō, Tadao. *Nihon eigashi, vol. 1.* Tokyo: Iwanami shoten, 1995.

Schacht, Roland. "Tonfilmdramaturgie." *Film-Kurier*, April 22, 1929.

Schickele, René. "Paneuropa der Sender" [1931]. In *Radio-Kultur in der Weimarer Republik*, edited by Irmela Schneider, 140–43. Tübingen: Narr, 1984.

Schweinitz, Jörg. "Wie im Kino! Die autothematische Welle im frühen Tonfilm. Figurationen des Selbstreflexiven." In: *Diesseits der 'Dämonischen Leinwand':*

Neue Perspektiven auf das späte Weimarer Kino, edited by Thomas Koebner, 373–92. München: edition text + kritik, 2003.

Seeber, Guido. *Der praktische Kameramann. Theorie und Praxis der kinematographischen Aufnahmetechnik*. Berlin: Verlag der Lichtbildbühne, 1927.

Seeber, Guido. *Der Trickfilm in seinen grundsätzlichen Möglichkeiten* [1927]. Frankfurt am Main: Deutsches Filmmuseum, 1979.

Seeber, Guido. *Kamera-Kurzweil. Allerlei interessante Möglichkeiten beim Knipsen und Kurbeln*. Stuttgart: Union Deutsche Verlagsgesellschaft, 1930.

Seldes, Gilbert. "Mickey-Mouse Maker," *The New Yorker*, December 19, 1931, 23–27.

Sellier, Geneviève. *Jean Grémillon: Le cinéma est à vous*. Paris: Klincksieck, 2012.

Seuffert, L. v. "Tonfilm-Impressionen." *Deutsche Filmzeitung*, January 30, 1930.

Sexton, Jamie. "The Audio-Visual Rhythms of Modernity: *Song of Ceylon*, Sound and Documentary Filmmaking." *Scope: An Online Journal of Film and Television Studies* 6, no. 2 (May 2004). Accessed February 2, 2023. https://www.nottingham.ac.uk/scope/documents/2004/may-2004/sexton.pdf.

Shibata, Kotaro. "'Realism' in late 1930s Japanese Films and Shiro Fukai's Accompaniment Music: Contemporary Arguments for the Rejection of Non-diegetic Music." *Aesthetics*, no. 21 (2018): 149–61.

Shigeno, Tatsuhiko. "Onkyō ni kansuru kansō." *Kinema junpō*, September 1, 1932.

Shimizu, Masao. "Jidaieiga shōron." *Kinema junpō*, September 11, 1932.

Shochiku, kabushikikaisha. "Shōchiku shichijūnenshi." Tokyo: Shochiku, 1964.

Simeon, Ennio. "Messter und die Musik des frühen Kinos." In *Oskar Messter. Filmpionier der Kaiserzeit,* edited by Martin Loiperdinger, 135–47. Frankfurt am Main: Stroemfeld, 1994.

Sisto, Antonella C. *Film Sound in Italy: Listening to the Screen*. New York: Palgrave Macmillan, 2014.

Slowik, Michael. *After the Silents: Hollywood Film Music in the Early Sound Era, 1926–1934*. New York: Columbia University Press, 2014.

Slowik, Michael. "Experiments in Early Sound Film Music: Strategies and Rerecording, 1928–1930." *American Music* 31, vol. 4 (2013): 450–74.

Smirnov, Andrey. *Sound in Z: Experiments in Sound and Electronic Music in Early 20th Century Russia*. London: Koenig Books, 2013.

"Sound Perspective Used in 'Applause." *Film Daily*, September 15, 1929, 10.

Spadoni, Robert. *Uncanny Bodies: The Coming of Sound Film and the Origins of the Horror Genre*. Berkeley: University of California Press, 2007.

Spergel, Mark. *Reinventing Reality: The Art and Life of Rouben Mamoulian*. Metuchen, NJ: The Scarecrow Press, Inc., 1993.

Spring, Katherine. *Saying It with Songs: Popular Music and the Coming of Sound to Hollywood Cinema*. New York: Oxford University Press, 2013.

Street, Sarah, and Joshua Yumibe. *Chromatic Modernity: Color, Cinema, and Media of the 1920s*. New York: Columbia University Press, 2019.

Sussex, Elisabeth. *The Rise and Fall of the British Documentary: The Story of the Film Movement Founded by John Grierson*. Berkeley: University of California Press, 1976.

Szczepanik, Petr. "The 'Birth' of Cinema History: The Coming of Sound Films and Media's Invention of Their Past." In *MLVs, Cinema and Other Media*, edited by Veronica Innocenti, 29–43. Udine: Campanotto Editore, 2006.

Szczepanik, Petr. "Sonic Imagination: or, Film Sound as a Discursive Construct in Czech Culture of the Transitional Period." In *Lowering the Boom: Critical Studies in Film Sound*, edited by Jay Beck and Tony Grajeda, 87–104. Urbana: University of Illinois Press, 2008.

Tamura, Yukihiko. "Madamu to nyōbō." *Kinema junpō*, August 21, 1931.

Tanaka, Masasumi, ed. *Shimizu Hiroshi: Sokkyōsuru poejī yomigaeru chō eiga densetsu*. Tokyo: Firumuatosha, 2000.

Taylor, Richard, and Ian Christie, eds. *The Film Factory: Russian and Soviet Cinema in Documents 1896–1939*. New York: Routledge, 1988.

Thiele, Wilhelm. "Operette im Tonfilm." *Film-Kurier*, February 1, 1930.

Thomasson, R. "Après trois ans d'absence, Charles Boyer revient parmi nous. Ses projets." *Pour Vous*, March 2, 1933.

Thompson, Kristin. "Early Sound Counterpoint." *Yale French Studies*, no. 60 (1980): 115–40.

Thorburn, David, and Henry Jenkins. *Rethinking Media Change: The Aesthetics of Transition*. Cambridge: MIT Press, 2004.

Toller, Ernst. "Internationaler Rundfunk" [1930]. In *Radio-Kultur in der Weimarer Republik*, edited by Irmela Schneider, 134–35. Tübingen: Narr, 1984.

Tomita, Mika, ed., *Chie puro jidai: Kataoka Chiezō Inagaki Hiroshi Itami Mansaku shadatsu ni entāteinmento*. Tokyo: Firumuatosha, 1997.

Tomoda, Junichirō. "Hachisuka Koroku daininen." *Kinema junpō*, September 11, 1929.

Tomonori, Saiki, ed. *Itō Daisuke: Hangyaku no passhon jidaigeki no modanizumu*. Tokyo: Firumuatosha, 1996.

"Tonfilme." *Deutsche Filmzeitung*, March 8, 1929.

"Tonfilmkritik. Äußerungen in der 'Deutschen Presse.'" *Deutsche Filmzeitung*, June 13, 1930.

Toshio, Iwase. "Nihon ni okeru toki kōgyo no mirai." *Eiga Ōrai* 6, no. 60 (February 1930): 72–75.

tp. "Zürcher Kinorundschau." *Neue Zürcher Zeitung*, February 7, 1930.

Tsekhanovskii, Mikhail. "Dela i liudi kak professional'nyi urok." *Proletkino*, no. 19/20 (1932): 15–16.

Tsivian, Yuri. "Cyberspace and its Precursors: Lintsbach, Warburg, Eisenstein." In *Transmedia Frictions: The Digital, the Arts, and the Humanities*, edited by Marsha Kinder and Tara McPherson, 80–99. Berkeley: University of California Press, 2014.
Tsuchihashi, Takeo. "Eiga ga koe wo hatatsushita toki." In *Kikigaki kinema no seishun*, edited by Kenji Iwamoto and Tomonori Saiki, 285–316. Tokyo: Riburopōto, 1988.
VanCour, Shawn. *Making Radio: Early Radio Production and the Rise of Modern Sound Culture*. New York: Oxford University Press, 2018.
Vial, Georges. "Le danger du cinéma parlant." *Le Courrier cinématographique*, July 21, 1928.
Vincendeau, Ginette. "Hollywood Babel: The Multiple Language Version." *Screen* 29, no. 2 (1988): 24–39.
"Voici les lauréats du jeu de la vedette la plus photogénique …" *Pour Vous*, December 17, 1931.
Volmar, Axel. "In Storms of Steel: The Soundscape of World War I and Its Impact on Auditory Media Culture during the Weimar Period." In *Sounds of Modern History: Auditory Cultures in 19th- and 20th-Century Europe*, edited by Daniel Morat, 227–55. New York: Berghahn, 2014.
Wadayama, Shigeru. "Chūshingura." *Kinema junpō*, January 1, 1933.
Wadayama, Shigeru. "Nikkatsu gendai eiga Haru to musume." *Kinema junpō*, June 21, 1932.
Wadayama Shigeru. "Toki no ujigami." *Kinema junpō*, April 21, 1932.
Wahl, Chris. *Multiple Language Versions Made in BABELsberg: Ufa's International Strategy, 1929–1939*. Amsterdam: Amsterdam University Press; Eye Filmmuseum. 2016.
Wahl, Lucien. "Imitateurs, attention!" *Pour Vous*, July 3, 1930.
Wedel, Michael. *Der deutsche Musikfilm: Archäologie eines Genres 1924–1945*. München: edition text + kritik, 2007.
Wedel, Michael. *Pictorial Affects, Senses of Rupture: On the Poetics and Culture of Popular German Cinema, 1910–1930*. Berlin: De Gruyter, 2019.
Widdis, Emma. "Making Sense without Speech: The Use of Silence in Early Soviet Sound Film." In *Sound, Speech, Music in Soviet and Post-Soviet Cinema* edited by Masha Salazkina and Lilya Kaganovsky, 100–116. Bloomington: Indiana University Press, 2013.
Weis, Elisabeth. *The Silent Scream: Alfred Hitchcock's Sound Track*. Rutherford: Fairleigh Dickinson University Press, 1982.
Weis, Elisabeth, and John Belton, eds. *Film Sound: Theory and Practice*. New York: Columbia University Press, 1985.
Wiegand, Daniel. "The Delightful Paradox: Towards an Aesthetics of Silence in Early Sound Film." In *Silence, Sounds, Music: Acoustic Dimensions of Immersion*,

edited by Peter Niedermüller, Florian Freitag and Laura Mücke. New York: Routledge, forthcoming.
Wiegand, Daniel. "'Eine Entdeckungsfahrt in die Welt der Geräusche'. Zur historischen Rezeption des frühen Tonfilms *Die Nacht gehört uns* (D 1929)." *Montage AV* 30, no. 2 (2021): 59–76.
Wiegand, Daniel. "'Islands of Sound in the Silent Flow of Film': German Part-Talkies around 1930 as a Hybrid Medium." *Historical Journal of Film, Radio, and Television* 24, no. 3 (2022): 427–50. https://doi.org/10.1080/01439685.2021.1976916.
Wiegand, Daniel. "Listening to Faint Sounds and Silence: Cinema's Transition to Sound and the Emergence of a New Auditory Sensitivity." In *Retuning the Screen: Sound Methods and the Aural Dimension of Film and Media History,* edited by Simone Dotto, 85–94. Milan: Mimesis International, 2022.
Williams, David. "Film and the Mechanization of Time in the Myth of the Great War Canon." *English Studies in Canada* 41, no. 2–3 (2015): 165–90.
Winkle, Ralph. "Phantasmagorien des Krieges – Kontinuität und Diskontinuität ästhetischer Muster in Filmen der dreißiger und vierziger Jahre." *Krieg und Literatur* 12 (2006): 369–81.
Wollenberg, Hans. "Der brave Sünder." *Die Lichtbild-Bühne,* October 23, 1931.
Wood, Nancy. "Towards a Semiotics of the Transition to Sound: Spatial and Temporal Codes." *Screen* 25, no. 3 (May–June 1984): 16–25.
Wood, Robin. *Howard Hawks.* New ed. Detroit: Wayne State University Press, 2006.
Wright, Basil and B. Vivian Braun. "Manifesto: Dialogue on Sound" [1934]. In *Film Sound: Theory and Practice,* edited by Elisabeth Weis and John Belton, 96–99. New York: Columbia University Press, 1985.
Wurtzler, Steve. *Electric Sounds: Technological Change and the Rise of Corporate Mass Media.* New York: Columbia University Press, 2007.
Yamamoto, Kikuo. "Rensai ronbun: hikaku eigashi kenkyu (26): Amerika eiga to jiyushugi jidai-geki, Inagaki Hiroshi kantoku sakuhin o megutte." *FC Firum senta,* no. 56 (1979): 30–36.
Yasar, Kerim. *Electrified Voices: How the Telephone, Phonograph, and Radio Shaped Modern Japan, 1868–1945.* New York: Columbia University Press, 2018.
Zagarrio, Vito. "The First Comedy, Italian Style: Blasetti, Camerini, De Sica." In *The Italian Cinema Book,* edited by Peter Bondanella, 58–65. London: British Film Institute, 2014.
"Zwei Prominente debütieren." *Mein Film,* no. 304 (1931): 3.
Zy. [= Victor Zwicky]. "Das lebende Bild." *Tages-Anzeiger,* February 8, 1930.
Zy. [= Victor Zwicky]. "Das lebende Bild." *Tages-Anzeiger,* October 25, 1930.

Index

For Product Safety Concerns and Information please contact our EU representative GPSR@taylorandfrancis.com
Taylor & Francis Verlag GmbH, Kaufingerstraße 24, 80331 München, Germany

www.ingramcontent.com/pod-product-compliance
Ingram Content Group UK Ltd.
Pitfield, Milton Keynes, MK11 3LW, UK
UKHW020016200926
475702UK00014B/1916
* 9 7 8 1 0 4 1 1 7 5 2 9 2 *